AN INTRODUCTION TO CLINICAL LABORATORY SCIENCE

An Introduction to Clinical Laboratory Science

JEANNE M. CLERC, ED.D., MT(ASCP)SH
Program Director
Clinical Laboratory Sciences Program
Eastern Michigan University
Ypsilanti, Michigan

St. Louis   Baltimore   Boston   Chicago   London   Philadelphia   Sydney   Toronto

**Mosby
Year Book**

Dedicated to Publishing Excellence

Sponsoring Editor: Stephanie Manning
Assistant Editor: Jane Petrash
Assistant Managing Editor, Text and Reference: George Mary Gardner
Production Project Coordinator: Maria Nevinger
Proofroom Manager: Barbara Kelly

Copyright © 1992 by Mosby–Year Book, Inc.
Mosby–Year Book, Inc.
11830 Westline Industrial Drive
St. Louis, MO 63146

All rights reserved. No part of this publication may be reproduced, stored in a retrieval system, or transmitted, in any form or by any means, electronic, mechanical, photocopying, recording, or otherwise, without prior written permission from the publisher. Printed in the United States of America.

Permission to photocopy or reproduce solely for internal or personal use is permitted for libraries or other users registered with the Copyright Clearance Center, provided that the base fee of $4.00 per chapter plus $.10 per page is paid directly to the Copyright Clearance Center, 21 Congress Street, Salem, MA 01970. This consent does not extend to other kinds of copying, such as copying for general distribution, for advertising or promotional purposes, for creating new collected works, or for resale.

2 3 4 5 6 7 8 9 0 CL/MV 96 95 94

Library of Congress Cataloging-in-Publication Data
Clerc, Jeanne M.
    An introduction to clinical laboratory science/Jeanne M. Clerc.
      p. cm.
    Includes bibliographical references and index.
    ISBN 0-8016-1392-2
    1. Medical laboratory technology—Vocational guidance. I. Title.
    [DNLM: 1. Career Choice. 2. Diagnosis, Laboratory.
3. Technology, Medical—education. 4. Technology, Medical—
-manpower. QY 21 C629i]
RB37.6.C57 1991   91-28015
616.07′5—dc20   CIP
DNLM/DLC
for Library of Congress

This book is dedicated to the memory of my father, ⸺⸺⸺⸺⸺⸺⸺⸺⸺⸺
Eugene H. (Jack) Clerc.

PREFACE

This book provides a description of clinical laboratory science (CLS). Many readers may not recognize this term, but are more familiar with the older term "medical technology." The American Society for Medical Technology (ASMT) and CLS educators encourage use of the new term. The text uses this new terminology primarily, and when appropriate lists synonymous words and terms.

The CLS profession faces new challenges in the 1990s. The shortage of qualified laboratory personnel is certainly one of them. Changes in the health care system over the past 10 years have been dramatic, and the future promises to bring more.

This text is primarily intended for college students who have chosen or are considering CLS as their field of study. Information is provided on career opportunities and educational requirements. In addition, it is critical that students entering the profession be groomed as professional practitioners as well as laboratory scientists.

With this in mind, ethics, communication skills, and commitment to professional organizations become important components of a CLS curriculum. Students also must have a good understanding of laboratory and health care organizations to function effectively in a rapidly changing environment. AIDS, federal and state governmental regulations, implementation of diagnosis-related groups (DRGs), and licensure issues all are "hot topics" for the 1990s, and are discussed in this text.

Related laboratory professions, such as cytotechnology, histotechnology, and nuclear medicine, also are considered clinical laboratory sciences. It is hoped this text will provide the type of information that will entice students to enter the clinical laboratory profession and the information essential for their becoming truly knowledgeable "health professionals."

Jeanne M. Clerc, Ed.D, MT(ASCP)SH

ACKNOWLEDGMENTS

As with most texts, the end product is the result of group effort. I would like to acknowledge a number of persons who contributed to this book.

I appreciate the efforts of two reviewers:

Jennifer Griffin, M.S., MT (ASCP)
Bronson Hospital, Kalamazoo, Michigan

Eleanor Wright, Ph.D.
Eastern Michigan University, Ypsilanti

and of contributing authors:

Susan Dingler, CT (ASCP)
Henry Ford Hospital, Detroit

Gary Hammerberg, Ed.D, MT (ASCP)
Eastern Michigan University, Ypsilanti

Clifford Renk, Ph.D.
Eastern Michigan University, Ypsilanti

Special thanks to our department secretary, Jane Bird, for assistance in preparing the manuscript. I also appreciate the personal and departmental support given me by my department head, Stephen Sonstein, Ph.D. Special thanks to the professional organizations that provided me with vital information, especially the American Society of Clinical Pathologists (ASCP), the American Society for Medical Technology (ASMT), and the National Certification Agency for Medical Laboratory Personnel (NCA).

Special thanks to two mentors: Dean Elizabeth King, Ph.D., and Dean Ronald Goldenberg, Ed.D., both at Eastern Michigan University. Their support and confidence in my abilities will be remembered always.

I am grateful for the sabbatical leave provided by Eastern Michigan University, which allowed me the time to complete this project.

Last, I would like to say thanks to my parents and family. Their love and support are deeply cherished.

Jeanne M. Clerc, Ed.D., MT(ASCP)SH

CONTENTS

Preface   *vii*

1 / Definition of Clinical Laboratory Science   *1*

    Nature of the Work   *1*
    Medical Technology or Clinical Laboratory Science?   *2*
    Scope of Practice   *3*
    Profession of Clinical Laboratory Science   *4*

2 / History of Clinical Laboratory Science   *5*

    Prior to 1990   *5*
    1900 to World War II   *6*
    Post–World War II   *6*

3 / Hospital and Laboratory Organization   *9*

    Classification of Hospitals   *10*
    Bureaucratic Principles   *11*
    Organization Chart   *12*
    Informal Organization   *14*
    Health Care Team Approach   *15*
    Professional Service Departments   *15*
    Clinical Laboratory Organization   *15*
    Physical Facilities   *19*
    Laboratory Organization: Past, Present, and Future   *19*

4 / Challenge of the 1990s: Is CLS for You?   *22*

    Intellectual Requirements   *22*
    Personal Characteristics and Emotional Requirements   *22*
    Physical Requirements   *23*
    Pros and Cons of the Profession   *23*
    Challenges of the 1990s   *25*

5 / Laboratory Staff, Salaries, and Advancement   *32*

    Laboratory Staff   *32*
    Salaries   *37*
    Advancement: The Career Ladder   *39*

6 / Registration, Certification, and Licensure   *44*

    Personnel Licensure   *45*
    Laboratory Licensure   *48*
    Federal Regulation of Laboratories   *49*

xii Contents

7 / Certification Agencies, Accrediting Agencies, and Professional Organizations 51

    Certification Agencies 51
    Independent Certification Agencies 54
    Accreditation Agencies 55
    Professional Organizations 57
    Information Material 62

8 / Overview of Laboratory Departments 64

    Coagulation Laboratory 64
    Hematology Laboratory 65
    Urinalysis Laboratory 66
    Blood Bank (Immunohematology Laboratory) 67
    Chemistry Laboratory 68
    Microbiology 71

9 / Educational and Clinical Requirements for Clinical Laboratory Technicians and Clinical Laboratory Scientists 73

    Clinical Laboratory Technician, Medical Laboratory Technician 73
    Clinical Laboratory Scientist, Medical Technologist 76
    CAHEA Program 79
    Applying to a Laboratory Science Program 81

10 / Clinical Laboratory Science Options 83

    Nuclear Medicine 83
    Cytogenetics 86
    Cytogenetic Technology Educational Programs: October 1989 89
    Cytotechnology 91
    Histology 94

11 / Employment and Career Opportunities 99

    Laboratory Technical Positions 101
    Outside Health Care 106
    Careers Requiring Further Education 110

12 / Professional Ethics 112

    Professionalism and Obligation 112
    Code of Ethics of the American Society for Medical Technology 112
    Definitions and Theories of Morality 113
    A Decision-Making Model 115
    An Application of the Decision-Making Model 115
    Using the Decision-Making Model 115
    Confidentiality 116
    AIDS and Ethics 117
    Risk and Obligation 117

Legal Issues and AIDS *118*
Situations: Ethics *119*

13 / Communication Skills and Public Relations *121*

Interpersonal Communication: A Model *122*
Defining Effective Communication *123*
Barriers to Communication *124*
Public Relations *124*
Confidentiality *125*
Improving Communication *126*
Nonverbal Communication *127*
Coping With Special Patients *128*
Telephone Manners *129*
Organizational Communication Systems *130*
Informal Communication Systems *130*

14 / Future of Clinical Laboratory Science *131*

Progress in Health Care *131*
The Laboratory of the Future *131*
Settings For Health Care Delivery *132*
Impact of New Technologies *133*
The Laboratory and Health Care Costs *134*
Personnel—Supply and Demand *134*
Laboratory Education *135*

Appendix A: Glossary *137*

Appendix B: Acronyms and Abbreviations *139*

Appendix C: Professional Organizations and Addresses *141*

Index *145*

Chapter 1

Definition of Clinical Laboratory Science

NATURE OF THE WORK

Medical technology, or clinical laboratory science (CLS), is a rapidly evolving industry of major proportions. In the United States alone, we spend some $20 billion each year for tests performed in more than 100,000 clinical laboratories.[3]

The practice of modern medicine would be impossible without the tests performed in the clinical laboratory. Although physicians depend heavily on laboratory results, ordinarily they do not perform the tests themselves. That is the job of clinical laboratory personnel.[7] A medical team of pathologists, clinical chemists, microbiologists, medical technologists, cytotechnologists, laboratory technicians, and assistants work together to determine the presence, extent, or absence of disease and provide data needed to evaluate the effectiveness of treatment.[2, 5]

Many clinical laboratories are highly automated, and job duties reflect this. Laboratory personnel work with a wide variety of specialized, high-precision instruments, including automatic analyzers, centrifuges, electronic counters, and computers. Many tests require the operation of complex electronic instruments to obtain such precise measurements as identifying quantities of certain hormones in the blood through the use of special radioactive chemicals. Clinical laboratory scientists work with all types of body tissues and fluids, from blood and urine to cell samples.

Clinical laboratory scientists may work in all general areas within the laboratory or may specialize in certain areas, for example, clinical chemistry, the determination of the presence and quantity of chemical substances in the blood; hematology, the specialization of abnormal conditions and diseases affecting the blood; microbiology, primarily concerned with detection of bacteria, fungi, viruses, and other organisms in the body; and immunohematology (blood banking), concerned with crossmatching and transfusing blood products. In these instances they may be known as chemistry, hematology, microbiology, or blood bank technologists.[6]

Because changes in body fluids, tissues, and cells often are a sign that something is wrong, clinical laboratory testing plays a crucial role in the detection and diagnosis of disease. Physicians also order laboratory work for other reasons. Test results may be used to establish values against which future measurements can be compared; to monitor treatment, as with tests for drug levels in the blood, which can indicate whether a patient is adhering to a prescribed drug regimen; to reassure patients that a disease is absent or under control; or to assess the status of a patient's health, as with cholesterol measurements.

Clinical laboratory scientists also monitor basic test results to ensure that each test is valid and that it meets high standards of precision, accuracy, and quality control.[6]

Depending on level of skill, the worker may run simple tests or perform complex analyses that require a number of steps to arrive at the information needed by the physician. The types of tests that clinical laboratory personnel perform and the amount of responsibility they assume vary with employment setting, but depend to a large extent on the kind of educational preparation they have.[7]

The profession is not bound by the four walls of the laboratory. Because the field is constantly changing, in response to new technologies, health care cost-containment pressures, and variations in health care needs, the nature of the work for laboratory scientists broadens each year.[3]

MEDICAL TECHNOLOGY OR CLINICAL LABORATORY SCIENCE?

Many definitions have been proposed for the term "medical technology." In simplest terms, medical technology is a *profession concerned with providing information based on the performance of analytical tests on human body substances to detect evidence of or prevent disease or impairment and to promote and monitor good health.*

This simple definition of medical technology focuses on the laboratory information function these practitioners provide, that is, on clinical laboratory data. A more current definition takes into account the important fact that since its origins the field has grown in complexity and responsibility from an ancillary, or supporting, occupation limited to generating test data to its status today as a diverse profession that includes roles other than clinical practice.[3]

In response, the terminology in laboratory science has changed. In the past, the narrow title Medical Technologist (MT) was used to denote a graduate of a baccalaureate program in laboratory science; now the title Clinical Laboratory Scientist is preferred. The literature and certain professional organizations still may refer to the field as Medical Technology (MT). The broader term "Clinical Laboratory Sciences" (CLS) more accurately reflects the career opportunities available in a variety of settings. CLS includes other laboratory personnel, such as histologists, histotechnologists, nuclear medicine technologists, cytogeneticists, and cytotechnologists.

Clinical laboratory scientists and other laboratory professionals are found throughout the health care delivery system, for example, in hospitals, independent commercial laboratories, clinics, physicians' offices, Red Cross and other blood banks, public health departments, and ambulatory care centers. Some clinical laboratory scientists work overseas, in the Peace Corps or Project Hope, or in government or private facilities in other countries.[3] Options outside the health care mainstream include biogenetics, occupational health, independent consulting, environmental health, industrial research, higher education administration and education, product development, marketing, sales, veterinary science, and criminology. The field of CLS offers baccalaureate graduates basic preparation in many areas.[3]

The field of clinical laboratory science is confusing to the public. The profession has low visibility within society and the health care system. Clinical laboratory professionals historically have had limited contact with patients. Most people know less about them than about other health care workers, such as nurses. Even so, the information the clinical laboratory professional provides is critical for the appropriateness of the care provided by nurses, family physicians, surgeons, pharmacists, and other health professionals. Each depends, in part, on accurate laboratory data to help plan or implement treatment and care for the individual patient.[3]

There are many reasons for the public's confusion about laboratory personnel. First, qualified laboratory personnel are known by a number of professional titles, for example, Clinical Laboratory Scientist (CLS) vs. Medical Technologist (MT), or Clinical Laboratory Technician (CLT) vs. Medical Laboratory Technician (MLT). Second, they work in a large number and variety of settings. Third, there is lack of consensus, even within the profession, as to what various clinical laboratory practitioners should be called and what their roles should be.[3]

Because of these overlapping titles and job descriptions, the task of counting laboratory personnel is an estimate at best. Estimated employment in 1986 of clinical laboratory technologists and technicians was 239,000. A 24% growth rate in employment from 1986 to 2000 is anticipated, which will mean an additional 57,000 personnel.[9]

SCOPE OF PRACTICE

The American Society for Medical Technology (ASMT), the oldest and largest of the professional societies devoted exclusively to clinical laboratory science, summarizes the scope of practice of the profession, which describes in general terms the services provided by clinical laboratory scientists. Clinical laboratory personnel, as members of the health care team, are responsible for[1]:

1. Assuring reliable test results which contribute to the prevention, diagnosis, prognosis, and treatment of physiologic and pathologic conditions. This assurance requires:
   a. Producing accurate test results.
   b. Correlating and interpreting test data.
   c. Assessing and improving existing laboratory test methods.
   d. Designing, evaluating, and implementing new methods.
2. Designing and implementing cost-effective administrative procedures for laboratories, including their services and personnel.
3. Designing, implementing, and evaluating processes for education and continued education of laboratory personnel.
4. Developing and monitoring a quality assurance system, to include:
   a. Quality control of services.
   b. Competence assurance of personnel.
5. Promoting an awareness and understanding of the services they render to the consumer/public and other health care professionals.

PROFESSION OF CLINICAL LABORATORY SCIENCE

Webster's New World Dictionary of the American Language[8] defines a profession as "a vocation or occupation requiring advanced training in some liberal art or science, and usually involving mental rather than manual work, such as teaching, engineering, writing, etc., especially medicine, law, or theology (formerly called the learned professions)."

Becoming a profession is a gradual process. Basic qualifications must be met to be considered an accepted, established profession. These qualifications include[4]:

1. Having a distinct body of knowledge in the discipline.
2. Maintaining standards of excellence.
3. Formulating a code of ethics.
4. Endeavoring to elevate the profession to a position of dignity and social standing.
5. Organizing and developing a professional, qualifying association.
6. Setting criteria for recruitment and training.

Tremendous gains have been made in each of these areas during the 1980s. Discussions on the doctorate as the terminal degree in clinical laboratory science indicates that there is a distinct body of knowledge. The ASMT strives to improve the image, in collaboration with other professional organizations. Now certification examinations are available through the National Certification Agency for Medical Laboratory Personnel (NCA), which is supported by ASMT. Both the American Society of Clinical Pathologists (ASCP) and ASMT have taken a proactive role in recruiting laboratory personnel. Further strides in these areas are anticipated in the 1990s.[4]

REFERENCES

1. Fiorella BJ, Maturen A: Statements of competence for practitioners in the clinical laboratory sciences. Am J Med Technol 1981; 47:649.
2. Gupta GC: *Allied health education directory 1990*, ed 18. Chicago, 1990, American Medical Association.
3. Karni K, Oliver JS: *Opportunities in Medical Technology Careers*. Lincolnwood, Ill, 1990, NTC Publishing Group.
4. Lindberg DS, Stevenson Britt M, Fisher FW: *Williams' Introduction to the Profession of Medical Technology*, ed 4. Philadelphia, 1984, Lea & Febiger.
5. Miller-Allan PA: *Introduction to the Health Professions*. Belmont, Calif, 1984, Wadsworth.
6. Nassif JZ: *Handbook of Health Careers: A Guide to Employment Opportunities*. New York, 1980, Human Sciences Press.
7. US Department of Labor: *Occupational Outlook Handbook*. Scottsdale, Ariz, 1988, Associated Book Publishers.
8. *Webster's New World Dictionary of the American Language*, ed 12. New York, 1968, World Publishing Co.
9. White MC: The 1988–89 job outlook in brief. *Occup Outlook Q* 1988; 32:29.

Chapter 2

History of Clinical Laboratory Science

The beginnings of the hospital clinical laboratory cannot be pinpointed. Prior to World War I there were few laboratory technicians. However, there were many individuals prior to World War I who helped lead the development of the modern clinical laboratory.

PRIOR TO 1990

Intestinal parasites such as *Taenia* and *Ascaris* were mentioned in writings dating to 1500 BC. Identification of intestinal parasites is still done today in the clinical laboratory.

Writings of Hippocrates (460–370 BC) indicate he knew of tuberculosis, malaria, mumps, and anthrax.

The study of urine sometimes helped in disease diagnosis during the Middle Ages (1096–1438 AD). It was found that certain urine samples attracted ants and that such urine had a sweet taste. It is likely these urine samples were from persons with diabetes mellitus.

A prominent physician at the University of Bologna in the 1300s employed Allesandra Giliani to perform certain tasks, including a few pioneer clinical laboratory tests.

Marcello Malpighi (1628–1694), a microscopist and anatomist, is considered by some the "father of pathology and histology."[7]

Antonj van Leeuwenhoek (1632–1723) is credited with the invention of the compound microscope. He described blood cells, was able to see protozoa, and classified bacteria.[7]

Developments in the 1800s included the establishment of cell theory and the development of organic chemistry, physiologic chemistry, and bacteriology. It was possible to grow bacteria, stain them, and study them under the microscope.

Dr. Douglas, at the University of Michigan, opened the first chemical laboratory and began laboratory instruction in 1844. The department of pathology was established there in 1850.

Rudolph Virchow (1821–1902) founded the Archives of Pathology in Berlin in 1847.

Hermann von Fehling (1812–1885), a German chemist, performed the first quantitative test for urine sugar in 1848.[4, 7]

Dr. William H. Welch taught the first laboratory course in pathology in an American medical school in 1878. He became the first professor of pathology at the Johns Hopkins University in 1885.[4]

The first clinical laboratory was opened at The Johns Hopkins Hospital, aided by donations from the Eaton estate, in 1896.[4]

1900 TO WORLD WAR II

The 1900 US census makes reference to 100 male technicians employed in the United States. The 1922 census indicated there were 3,500 technicians by 1920, the majority being women. These numbers included dental and industrial workers as well as laboratory technicians. Physicians were still doing most of their own laboratory work.[4, 5]

A Manual of Clinical Diagnosis was published by James C. Todd[8] in 1908. Later editions were a joint effort of Dr. Todd and Dr. Arthur Sanford; more than 17 editions of that book, now entitled *Clinical Diagnosis and Management by Laboratory Methods*, have been published.

George Papanicolaou (1862–1962), a US scientist, described the Pap smear, or Pap test, in which cells collected from the cervix and vagina are examined to detect cancer cells.[3, 7] He helped train pathologist assistants, forerunners to laboratory technicians and technologists, in an apprentice setting.[3]

About the time of World War I the typical hospital pathology department was small and ill-equipped. Pathologists operating the department spent most of their time doing relatively simple blood cell counts, urinalyses, and a few chemical determinations. There were few surgical procedures. Tissue analyses and autopsies were rare. In the 1920s, pathologists and the clinical laboratories played a more major role in health care. Pathologists learned much from treating the wounded of World War I, and they applied these techniques in civilian hospitals. Great scientific discoveries were adopted in the hospital clinical laboratories in a short time. During this period more medical journals were beginning to be published.[6, 10]

In 1923 one of the first schools to train laboratory workers was instituted at the University of Minnesota. The school still exists.

During the 1920s and 1930s state societies of clinical pathologists were formed. By 1936 pathologists had established the American Board of Pathology, defined the requirements of the specialty, and were recognized by the American Medical Association (AMA).

In 1937 blood banking became a practical procedure in American hospitals. Blood banking is now an essential service in a hospital clinical laboratory. The procedures of compatibility testing and preparation and storage of blood are performed in the blood bank.[6, 8]

POST–WORLD WAR II

The technology of medicine that developed during World War II had a marked effect on laboratory medicine. Pathology and clinical laboratories continued to grow after World War II. New knowledge of health and disease problems proliferated. Antimicrobial agents were in general use. Rapid advances in instrumentation produced the most visible effect. Diagnostic

radioisotopes, exfoliative cytology, molecular biology, practical virology, and fluo
ies were introduced.[4, 6, 8]

In the 1950s almost all clinical chemistry procedures were performed n chemistry bench contained an assortment of reagent bottles. The technologist, usually, woman, would assemble a rack of specimens and several racks of empty test tubes. A sectioned drawer in the bench held a collection of pipets.

To process a specimen, the technologist reached for a clean pipet and transferred the appropriate amount of serum to a clean test tube. She often made use of a boiling water bath heated by a Bunsen burner. She then moved to a colorimeter or spectrophotometer and took a reading for each specimen. Then she would record the readings on a log sheet, perform the necessary calculations, and note the results on the log sheet and the patient's report.[1]

Today technologists usually set up an instrument for an entire shift or day, transfer unmeasured specimens into special containers, inspect printed results, record quality control data, and perform routine instrument maintenance.[1]

The introduction of quality control concepts and procedures to the clinical laboratory is one of the major achievements of the past 40 years. Levey and Jennings introduced their quality control chart to clinical chemistry in 1950. In 1958 Freier and Rausch presented the first comprehensive daily quality control program for laboratories. Later quality control methods were improved and broadened into the concept of quality assurance, dealing with reporting, data management, and the use of tests and data.[2]

Chromatography was discovered by Tswett in 1906, and led to the development of gas-liquid chromatography by Martin and James in 1952.[2]

Ouchterlony's radial immunodiffusion method in gel was described in 1949. Grabar and Williams used some of these principles to develop immunoelectrophoresis in 1953. In 1966 Laurell introduced "rocket" immunoelectrophoresis.

In 1957, Technicon introduced the first commercial continuous-flow automatic analyzer, the AutoAnalyzer, designed by Leonard Skeggs. This allowed for reliable, rapid analysis of several blood constitutents within a reasonable amount of time. Larger multichannel automated instruments soon followed. Screening batteries became part of the admission examination in hospitals and clinics.[2]

The first radioimmunoassay method was developed based on the 1960 Berson-Yalow studies on the state of ^{131}I-insulin in plasma. With radioimmunoassay, concentrations of substances such as hormones, pharmaceutical agents, and vitamins could be determined. Better understanding of diabetes mellitus soon followed.

Other advances in laboratory medicine in the 1950s and 1960s included absorption spectrophotometry, flame photometry, and atomic absorption. The field of cytogenetics was developed, and several major chromosomal abnormalities were discovered, including the Philadelphia chromosome found in 1960 by Nowell and Hungerford in chronic myelogenous leukemia. The major histocompatibility complex (the HLA system) was uncovered by Dausset and co-workers in 1965, aiding in successful organ transplants.

The 1960s marked the beginning of government interest and participation in health care through federal legislation and the advent of Medicare and Medicaid. The nationwide introduction of Diagnosis Related Groups (DRGs) in 1983 imposed a new cost consciousness on medicine, the hospital industry, and clinical laboratories.

In the 1970s and 1980s, automated instruments allowed for discrete sampling and random access, providing much more flexibility, versatility, durability, and direct on-line computer operation. Computers moved into the clinical laboratory in 1969, and proliferated during the 1970s and early 1980s, chiefly because of advances in technology.

Ion-selective electrodes began to appear in the clinical laboratory in the early 1970s. Na^+ and K^+ electrodes followed in the later 1970s.

A significant breakthrough was the introduction of hybridoma technology by Kohler and Milstein in 1975, which permitted production of monoclonal antibodies. This advance has had immense impact on the fields of immunochemistry and immunology, leading to the discovery of "tumor markers."

The growth of technology and knowledge over the past 40 years is remarkable. The 1990s promise to equal the exciting advances of the 1970s and 1980s.[2]

REFERENCES

1. Baer DM: Redesigning the lab for today's workflow. *Med Lab Observ* 1989; 21:26.
2. Benson E: The past as prologue: A look at the last 20 years. *Med Lab Observ* 1989; 21:27.
3. Gupta GC: *Allied Health Education Directory 1990*, ed 18. Chicago, 1990, American Medical Association.
4. Lindberg DS, Stevenson Britt M, Fisher FW: *Williams' Introduction to the Profession of Medical Technology*, ed 4. Philadelphia, 1984, Lea & Febiger.
5. Miller-Allan PA: *Introduction to the Health Professions*, Monterey, Calif, 1984, Wadsworth.
6. Snook ID: *Hospitals: What They Are and How They Work*. Rockville, Md, 1981, Aspen Systems Corp.
7. Thomas CL: *Taber's Cyclopedic Medical Dictionary*, ed 16. Philadelphia, 1989, FA Davis.
8. Todd JC: *A Manual of Clinical Diagnosis*. Philadelphia, WB Saunders, 1908.
9. Todd JC, Sanford A: *Clinical Diagnosis and Management by Laboratory Methods*, ed 17. Philadelphia, WB Saunders, 1984.
10. Wolper LF, Pena JJ: *Health Care Administration*. Rockville, Md, 1987, Aspen Systems Corp.

Chapter 3

Hospital and Laboratory Organization

THE U.S. HEALTH CARE SYSTEM

Although there are many career opportunities for laboratory professionals, a majority of clinical laboratory scientists and clinical laboratory technicians continue to be employed at hospitals within the health care system.

Visitors from abroad often ask about the American health care system and how it works. They are usually puzzled by the answer they get: There is no single American health care system. Many separate subsystems serve different populations in different ways. Sometimes these subsystems overlap; sometimes they stand apart from one another.[9]

In the earliest days, health care was entirely private and people were expected to take care of themselves by obtaining services of private physicians and nurses when needed and purchasing medications from drugstores and chemist shops. Charitable institutions were established as voluntary, nonprofit corporations to provide health care for those who could not obtain it on their own. These groups usually centered their efforts around hospitals and usually were located in larger towns and cities.

As the cost of health care began to increase rapidly after World War II, a wide variety of health insurance plans became available in the United States. The first to be operated were community-based, nonprofit, Blue Cross and Blue Shield plans, developed by hospital and physician associations to spread the cost of health care more widely among the population.

Private medical practitioners, voluntary nonprofit hospitals, city and state government hospitals, military and veterans hospitals, and health insurance plans all were developed at the same time, separately and for specific purposes.[9]

In 1988 expenditures for health care in the United States totaled more than $539.9 billion, nearly 12% of the nation's gross national product (GNP), an all-time high. Estimated spending by the year 2000 will be approximately 15% of the GNP. It is estimated that $16 billion of the nation's total health care expenditures is for clinical laboratory services. About 7,000 hospital laboratories and thousands of physicians' office laboratories perform most analyses, about 75%; the remaining 25% are performed in approximately 6,500 independent laboratories.[6, 8]

CLASSIFICATION OF HOSPITALS

The hospital industry is complex and diverse, and therefore difficult to describe simply. However, hospitals can be classified on the basis of a variety of characteristics.

Public Access.—One of the oldest and most useful systems of classification is used by the American Hospital Association (AHA) in its annual *Hospital Statistics* manual. That classification divides hospitals into community and noncommunity hospitals. Community hospitals include all nonfederal short-term general and other special hospitals, excluding hospital units of institutions, with facilities and services available to the public.

Noncommunity hospitals include federal hospitals, hospitals providing long-term care, and hospital units in institutions, such as prison hospitals or college infirmaries. The largest portion of the noncommunity hospitals comprises the 341 federal hospitals, with a bed capacity of 112,308, which in 1985 accounted for 3,073,885 admissions. The second largest group includes the 267 state and local psychiatric hospitals, with 145,980 beds and 327,521 admissions in 1985.[10]

Ownership.—Another classification is based on ownership, that is, government or public ownership or control of the policies and operation of the hospital. Institutions are divided into four groups under this classification: government, nonfederal; nongovernment, not for profit (religious or voluntary); investor owned, for profit; and government, federal.[9, 10]

Duration of Patient Stay.—The most common type of hospital is the short-stay or short-term hospital, where most patients have acute conditions requiring hospital stays of less than 30 days. The average duration of stay in short-term hospitals is less than 7 days. In long-term hospitals, most of which house patients with chronic disease, psychiatric disorders, or tuberculosis, the average duration of stay is 3 to 6 months. According to AHA statistics, in 1985 there were 5,843 hospitals listed as short-term facilities and 528 as long-term hospitals; most (377) of the long-term facilities were psychiatric hospitals.[9, 10]

Number of Beds.—Hospitals are grouped by number of beds: 6 to 24 beds, 25 to 49, 50 to 99, 100 to 199, 200 to 299, 300 to 399, 400 to 499, and 500 or more. This categorization usually is combined with other classifications, such as regional and teaching or nonteaching hospital.

Accreditation.—Hospitals are classified as accredited and nonaccredited, depending on whether they have been found in substantial compliance with the standards of the Joint Commission on Accreditation of Healthcare Organizations and/or the American Osteopathic Association.

Teaching.—Teaching hospitals participate in the education of physicians through a residency program. Depending on the type and number of residency programs offered, a hospital is either a major or minor teaching institution.[10]

Type of Service.—Hospitals are classified by the type of service provided in the majority of admissions. The four major categories are general medical and surgical, psychiatry, tuberculosis, and other specialty services. Predominant is the general hospital offering a wide range of medical, surgical, obstetric, and pediatric services. Specialty hospitals provide care for a specific disease or population group.[9, 10]

Vertical Integration.—Hospitals can be classified according to vertical integration, or "regionalization." Under this system, hospitals are categorized as primary care, secondary care, and tertiary care centers.

Primary care facilities offer services on a need/demand basis to the public. They operate as an essential component of a comprehensive health care system and provide health services in an available and continuous fashion on an outpatient basis.

Secondary care facilities render care that requires a degree of sophistication and skills, and are usually associated with confinement of the patient for a definite period of time. General acute hospitals or specialized outpatient facilities, such as ambulatory surgical centers, are included in this category.

Tertiary care facilities render highly specialized services requiring highly technical resources. This type of care is usually offered by university medical centers or specialty hospitals, such as burn centers.[10]

BUREAUCRATIC PRINCIPLES

Most hospitals in the United States tend to be traditionally organized; that is, they follow the classic theory of organization. The traditional organization structure derives from the theory of bureaucracy developed by the 19th century German sociologist Max Weber (1864–1920) and others. These bureaucratic principles include the following.

Division and Specialization of Labor.—Written job descriptions and task lists are mandatory tools for the modern-day hospital.

Pyramidal Organization or Scalar Principle.—At the top of the pyramid is a department head who has clearly specified lines of authority. This authority is passed down to employees at the lower levels in the hierarchy. These levels of authority create another principle, the "chain of command."[6, 10]

System of Rules and Regulations.—Rules and regulations could include such procedural issues as completing purchase requisitions, personnel regulations, and addressing scheduling patterns.

Unity of Command.—Each employee in the hospital is responsible to one person. This traditional approach of a single boss is part of the concept of the chain of command and authority.

Span of Control.—Each supervisor can properly direct only a limited number of subordinates or functions. Depending on the institution, most organization charts fit one of two basic designs: flat organizational structure, with few levels of hierarchy (Fig 3–1, A), or tall organi-

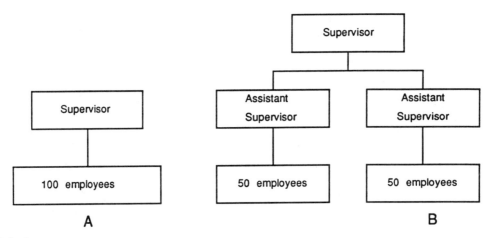

FIG 3–1.
Span of supervision. **A,** flat structure. **B,** tall structure. (From McBride KB: *Textbook of Clinical Laboratory Supervision*. New York, Appleton-Century-Crofts, 1982. Used by permission.)

zational structure, with numerous levels (Fig 3–1, B). The tall structure generally is used when there is increased specialization; the flat structure is characterized by a broad span of control, that is, many subordinates responsible to one superior. Classic organization theories hold that the more efficient organizations have a smaller span of control.[2, 6, 10]

Delegation.—Delegation can be defined as the transfer of authority to another person to accomplish a specific finite goal under carefully defined conditions.[6]

Line and Staff.—Line authority in the hospital connotes direct supervision over subordinates in a direct authority relationship. For example, the head nurse is directly responsible for the work of employees under his or her supervision. The staff function is associated with advisory activities rather than direct supervision. An example might be an educational component, consisting of trainers and educators, for inservice education.

Coordination.—Coordination helps ensure that all work efforts within the institution are synchronized.

Project Management.—Hospital managers often form project groups, task forces, or teams.

ORGANIZATION CHART

The management tool for depicting organizational relationships is the organization chart, a visual arrangement that details the following aspects of an institution[3]:

1. Major functions, usually by department.
2. Respective relationships between functions or departments.

Hospital and Laboratory Organization   **13**

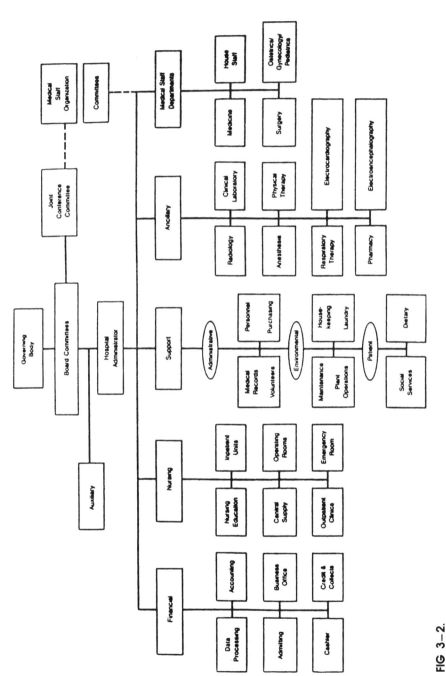

FIG 3–2.
Typical hospital organization chart. (From Snook ID Jr: *Hospitals: What They Are and How They Work*. Rockville, Md, Aspen Systems Corp, 1981, p 19. Used by permission.)

3. Channels of supervision.
4. Lines of authority and of communication.
5. Positions (by job title) within departments or units.

The organization chart attempts to show relationships between line and staff. In this organizational concept, a *line position* is one in which a superior exercises direct supervision over a subordinate. Supervisory relationships are indicated by vertical placement and solid lines (e.g., in Fig 3–2, the relationship of the board committees to the Hospital Administrator). A *staff position* is advisory, supportive, or auxiliary, indicated by equal placement vertically on the chart and connected by a broken line (e.g., the Joint Conference Committee and Medical Staff Organization in Fig 3–2).

A typical hospital organization chart is seen in Figure 3–2. The organization chart not only identifies Chief Executive Officer (CEO) functions, it also is a guide for the hospital family, delineating the sphere of influence for the middle managers.

At the top of the hospital organization is a governing body, that is, a board of trustees, board of directors, or board of governors. The board is headed by a chairman who holds ultimate responsibility for hospital operations. The board typically functions through committees, which are established for key functions such as finance, long-range planning, fund raising, and quality assurance. The board is responsible for medical staff actions, and hires the administrator. Hospital trustees often are private citizens, serve without pay, and are prohibited from profiting in any way from membership on the board of trustees.[5]

At the head of the hospital medical team is either a chief of staff, a medical director, or a president of the medical staff. The role of the chief of staff or president of the medical staff is to serve as liaison between the medical staff and its physicians, the board of trustees, and the hospital administration, the team of *three players*.[9,10]

The third member of the hospital team is the hospital administrator, or CEO. This person may have the title of president, executive director, or hospital administrator. The CEO is responsible for all matters relating to the operation of the hospital.

The activities of each part of the triad have significant impact on the others. The board of trustees delegates its authority to the hospital CEO, who then has the responsibility to develop the hospital management organization. The CEO has flexibility in structuring the management tasks and organization and defining the management principles under which the hospital will operate. Assistant administrators generally assist the CEO in a variety of functions, such as Financial, Nursing, Support, Ancillary, and Medical Staff Departments (*see* Fig 3–2).

In the long run, no hospital can be successful without effective communication and cooperation between the board of trustees, the CEO, and the medical staff.

INFORMAL ORGANIZATION

One of the severe limitations of a hospital's formal organization chart is that it does not reflect the hospital's informal organization. The organizational hierarchy, no matter how well designed, does not ensure total organization. It is not always true that orders flow down from the manager and that feedback information flows up from the subordinates to the manager. Lines of communication in the informal structure may be substantially different from those planned in the formal structure. Often information is communicated through informal channels rather than through channels delineated in the formal structure. An example is the CEO who during a social lunch with the laboratory manager passes along information that might not be communicated through a formal communication process.[6, 10]

HEALTH CARE TEAM APPROACH

For many years there were only two health professions: physician and nurse. The technologic progress of the last century, and especially since 1940, has been explosive. In less than a hundred years, dozens of specialized health occupations have come into being, and more are developed every year.

Many differences exist between and among professionals and nonprofessionals in these health occupations. It falls to the manager to ensure that the contributions of all of these perhaps diverse persons are pooled and effectively focused toward the care of the organization's patients. A variety of health professionals will be involved in a particular patient's care. For example, a patient with a respiratory tract infection may receive respiratory treatment from a respiratory therapist, have diagnostic tests completed through radiology and laboratory services, and receive antibiotic therapy through the pharmacy, in addition to the standard physician and nursing care.[4]

PROFESSIONAL SERVICE DEPARTMENTS

The professional service departments are also referred to as ancillary departments, or those departments that assist the physician and medical team in diagnosing and treating disease. Prior to 1983 these ancillary departments were considered to be revenue centers; that is, they billed either the patient directly or the patient's insurance carrier for services. These professional service departments include the Clinical Laboratory, Radiology (x-ray), Anesthesiology, Physical Therapy, Inhalation Therapy, Electrocardiography Unit (heart stations), Electroencephalography, and Pharmacy.

In 1983, reimbursement procedures for Medicare admissions were changed from reimbursement based on the actual costs of services provided to payment based on the diagnosis (diagnosis-related group, DRG). Thus ancillary services are now considered cost centers rather than revenue-generating centers.[10]

CLINICAL LABORATORY ORGANIZATION

Figure 3–2 can be considered a master chart, depicting the entire organization, although not in great detail. Figures 3–3 and 3–4 are supplementary charts of a department, specifically, the hospital laboratory.[3]

Figure 3–3 is a Pathology organization chart. When a laboratory manager speaks of staff, he or she generally is referring to technical personnel serving in line positions. Individuals involved solely in academic instruction or research are defined as staff to the functioning ana-

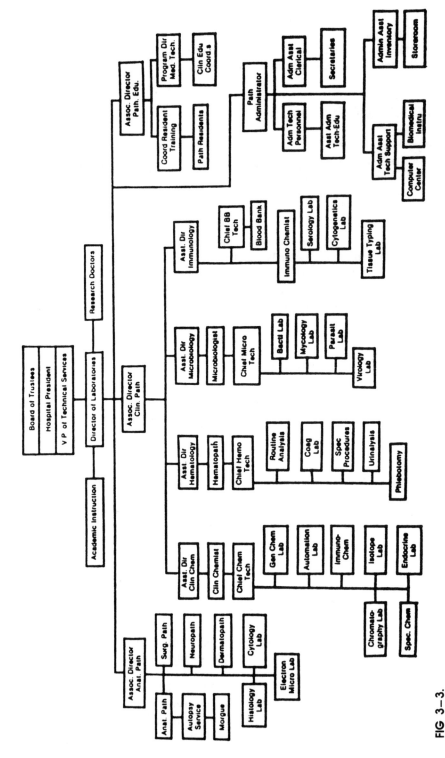

FIG 3–3.
Pathology organization chart. (From Snyder JR, Senhauser DA: *Administration and Supervision in Laboratory Medicine*, ed 2. New York, JB Lippincott, 1989, p 35. Used by permission.)

tomic and clinical laboratories. Another component of the structure, more visibly staff, is the administrative section. Its role is supportive to the primary laboratory testing function.[6]

The department itself can be divided into two major sections, a clinical pathology division and an anatomic pathology division. The clinical laboratory and pathology services fall into two areas: those performed directly by the pathologist and those performed under the pathologist's responsibility and supervision but actually conducted by other laboratory personnel.

The department is organized under the leadership and directorship of a *pathologist*, a licensed physician who specializes in the practice of pathology and is usually eligible for certification or is certified by the American Board of Pathology in either clinical or anatomic pathology, or both. Usually one pathologist is designated as the director. Large teaching/research facilities may appoint one director of anatomic pathology (AP) and another director of clinical pathology (CP).[5, 10]

The anatomic pathologist examines all surgical specimens (including autopsy specimens, frozen sections, Pap smears, and tissue consultations). The cytology and histology laboratories generally are under the supervision of an anatomic pathologist. Occasionally in large teaching centers electron microscopy may be assigned to the anatomic division. The roles of the *histotechnologist* and *cytotechnologist* are discussed more fully in Chapter 10.

The clinical pathology division processes the more common laboratory test requests, including blood cell counts, coagulation studies, urinalysis, blood sugar determinations, and throat cultures. Its subsections include hematology, chemistry, blood banking, microbiology, immunology, and urinalysis, among others. Again, larger hospitals may expand these sections to include more defined units, such as a coagulation section (usually incorporated in hematology). Other subsections may include endocrinology, enzymology, serology, and tissue typing laboratory. The clinical pathologist assists the primary care physician as a consultant, when requested, to select tests and interpret laboratory results.

To staff these areas clinical laboratories employ a blend of medical, technical, and support staff. The number of employees, titles, and job descriptions varies according to the size of the institution and the scope of services provided. Figure 3–4 is an organizational staffing chart for a typical medium to large hospital laboratory. Most hospitals employ people in the positions of Pathologist, Laboratory Scientist/Specialist, Section Supervisor, Technologist, Technician, Phlebotomist, and Clerk/Medical Secretary.

Laboratory Scientists/Specialists may be employed to assist pathologists. These individuals have advanced degrees, usually doctorates, in areas such as biochemistry, microbiology, or virology. They usually are found in large teaching centers, and their responsibilities generally include a large research component. Specialists also include those laboratory scientists who have completed additional education and passed a specialist examination, such as Specialist in Blood Banking (SBB).

Laboratory Managers/Administrative Technologists/Laboratory Coordinators usually are medical technologists with administrative skills acquired through experience or formal education (e.g., an M.B.A. program). In very large institutions they may have only a business background with no clinical experience. In this case, one of two staffing patterns usually is found. A chief technologist is responsible for the technical aspects of managing the department, including quality control, instrument maintenance, and finances. The *Section Supervisors* assume responsibility for technical supervision of a section of the laboratory, for example, chemistry or hematology. Their time is split between technical work (test analyses) and supervision. In smaller hospitals one supervisor may be responsible for several sections.[10]

18 Chapter 3

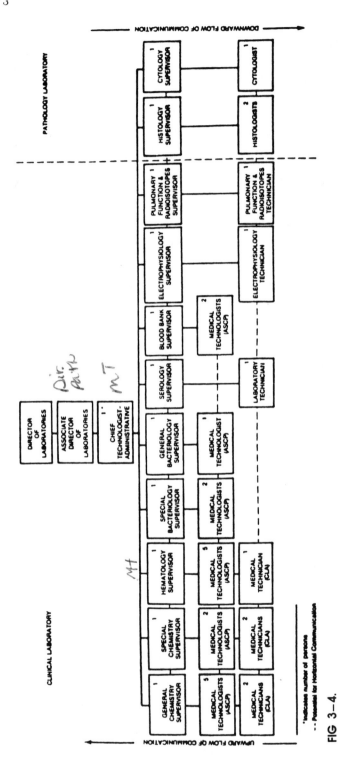

FIG 3-4.
Example of a clinical laboratory/pathology laboratory organization chart. (From Snyder JR, Senhauser DA: Administration and Supervision in Laboratory Medicine, ed 2. New York, JB Lippincott, 1989, p 75. Used by permission.)

In larger institutions, where one section may employ as many as 30 or 40 staff technologists, section supervisors do limited, if any, bench work. The role of a section supervisor is determined by the size and scope of the operation. *Senior Technologists* may assist the section supervisor in major subsections of the department. For example, in the microbiology department of a large hospital senior technologists may be assigned to subsections such as mycology, virology, parasitology, or anaerobes.[10]

Medical Technologists (MT), or *Clinical Laboratory Scientists* (CLS), often possess a bachelor of science degree in medical technology or clinical laboratory science(s). *Medical Laboratory Technicians* (MLT), or *Clinical Laboratory Technicians* (CLT), possess less formal education, perhaps an associate degree. Laboratory professionals prefer the titles CLS and CLT, believing these terms are more descriptive of the profession and job responsibilities.

The *Phlebotomist's* primary job is to procure blood specimens. *Clerks/Secretaries* process information in the laboratory, sort patient reports, register outpatients, do general typing, and handle departmental mail.

Other positions may include Education Coordinator, Quality Control Supervisor, and Information Supervisor. These jobs, usually found in large institutions, require the educational background of a clinical laboratory scientist, that is, a B.S. degree.[10]

A more complete description of laboratory staffing and the career ladder is presented in Chapter 4.

PHYSICAL FACILITIES

The physical facility also contributes to the laboratory organization. Factors in designing the laboratory include workload, staffing, service levels, equipment, and work flow.

Medical laboratories follow two basic structural designs: modular and open. A modular laboratory emphasizes departmentalization, with a separate room for each laboratory division. In an open laboratory design many of the departments interface freely without dividing walls.[6]

LABORATORY ORGANIZATION: PAST, PRESENT, AND FUTURE

Laboratories must face a growing shortage of qualified personnel just when the "graying of America" indicates there will be an increased demand for laboratory tests. The challenges of the 1990s dictate a long, hard look at the traditional organizational structure of the U.S. laboratory.

Changes in the laboratories from the 1950s to 1980s include an increase in middle management positions, such as department supervisors. With an increase in middle management positions, bureaucracy grows. Figure 3–5 is an organization chart for the future, the 1990s and beyond. The setup differs greatly from earlier laboratory structures.[1]

The organization chart of the future suggests that the ranks of management personnel will be markedly smaller. Sections have been consolidated. Former section supervisor slots have

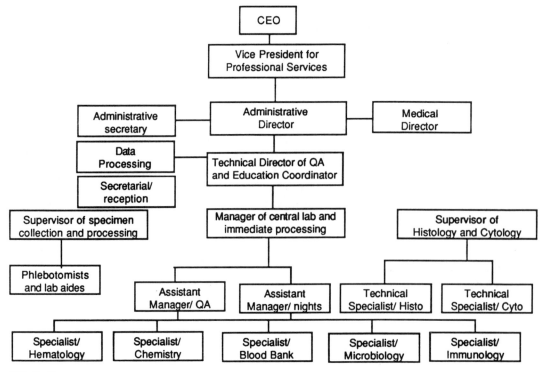

FIG 3–5.
Laboratory of the 1990s. (From Barros A: *MLO*, Dec 1989, p 37. Used by permission.)

been reclassified as Technical Specialist. The title has changed, but the job responsibilities remain the same. The specialist title indicates the reality that section supervisors spend time at the bench.

The reduction in supervisory slots delegates administrative functions (e.g., budgeting, scheduling, employee counseling) to a few individuals. Section consolidation makes staff rotation easier to manage.[7]

The dual promotional ladder created by this plan recognizes technical expertise by means of the "technical specialist" designation. The manager and general supervisor slots stress management skills.

Although technical specialists may or may not have specialist certification, they will be expected to prove their competence and knowledge in their chosen disciplines.

The manager or supervisor may not have the depth of technical expertise but must demonstrate managerial skills.

The organization chart of the 1990s shows a central laboratory that utilizes an immediate processing unit. There are five clinical laboratory disciplines, along with afternoon and evening shifts. The "stat" lab likely will be one section of the main lab, for efficiency. "Stat" laboratory tests must be performed immediately, in contrast to most laboratory tests, designated as "routine," in which time is not such a critical factor.[1]

REFERENCES

1. Barros A: Reorganize your lab structure for productivity in the '90s. *Med Lab Observ* 1989; 21:33.
2. Becan-McBride K: *Textbook of Clinical Laboratory Supervision.* New York, 1982, Appleton-Century-Crofts.
3. Liebler JG, Levine RE, Dervitz HL: *Management Principles for Health Professionals.* Rockville, Md, 1984, Aspen Systems Corp.
4. McConnell CR: *Managing the Health Care Professional.* Rockville, Md, 1984, Aspen Systems Corp.
5. Snook ID: *Hospitals: What They Are and How They Work.* Rockville, Md, 1981, Aspen Systems Corp.
6. Snyder JR, Senhauser DA: *Administration and Supervision in Laboratory Medicine*, ed 2. Philadelphia, JB Lippincott, 1989.
7. Umiker WO: Should labs eliminate supervisors' jobs? *Med Lab Observ* 1988; 20:37.
8. Van der Vaart D: Health care spending on increase again. *Advance* 1990; 3:18.
9. Williams SJ, Torrens PR: *Introduction to Health Services.* New York, John Wiley, 1980.
10. Wolper LF, Pena JJ: *Health Care Administration.* Rockville, Md, 1987, Aspen Systems Corp.

Chapter 4

Challenge of the 1990s: Is CLS for You?

You may have already committed yourself to a clinical laboratory science (CLS) major or you may be undecided and wish to explore other professions. How do you know if CLS is for you? Consider two things:

1. Do you have the personality traits and skills that suggest you would be fitted for this profession?
2. Would you enjoy the types of tasks and responsibilities assigned to a clinical laboratory scientist? For example, if you do not like high school or college science laboratories, explore other professions or choose an alternate career path in CLS.

INTELLECTUAL REQUIREMENTS

Clinical laboratory science majors should be good students, intelligent, hard-working, and motivated. The CLS science-based profession requires considerable intelligence and problem-solving abilities. Interest and aptitude in biology are essential, and competence in mathematics and chemistry is necessary. The CLS curriculum is demanding, and requires such courses as anatomy, physiology, biologic science, biochemistry, and microbiology.[13]

PERSONAL CHARACTERISTICS AND EMOTIONAL REQUIREMENTS

There is an association between personality, vocation, and job satisfaction. The research to date indicates that persons attracted to and remaining in the laboratory profession have a consistent set of personality traits. Satisfied clinical laboratory scientists are generally realistic, practical, observant, and good at working with and remembering facts. They prefer to work in a planned and orderly manner based on facts and experience, are adaptable to routine, and are somewhat independent.[19]

Accuracy, dependability, and the ability to work under pressure are important personal characteristics for clinical laboratory personnel. Close attention to detail and accuracy is essential because small differences or changes in test substances or numerical readouts can be critically important for patient care. With the widespread use of automated laboratory equipment, mechanical, electronic, and computer skills are gaining importance. In addition, clinical laboratory scientists are expected to be good at problem solving and to have strong interpersonal and communications skills.[21]

Clinical laboratory scientists must be able to prioritize. Some items require immediate action, such as a "stat" laboratory test. There may be times when two or three tasks are critical. Clinical laboratory scientists must be able to make judgments, organize the work load, and carry out tasks with little supervision. Successful clinical laboratory scientists are curious, are able to follow detailed written and oral instruction, enjoy learning new things, and are diligent; they stick to the task.[13]

Clinical laboratory scientists perform volumes of precision work under considerable pressure; life-and-death situations often join productivity demands to escalate the pressures under which they perform. They must have a strong desire to help others. These practitioners must also have an extremely high level of integrity and personal responsibility. Patients' lives depend on their commitment to provide the highest quality service. These practitioners must also be emotionally stable and patient.[21]

Stress does affect work performance, health, and the quality of life. Stress exacts a heavy toll in absenteeism, turnover, health costs, error, and accident rates. Particularly affected is the field of clinical laboratory science. Of 130 occupations considered, laboratory professionals ranked seventh highest in work-related stress.[11]

Laboratory testing demands extremely high levels of accuracy in performance, frequently under severe time constraints, with the consequences of an error almost always being serious and possibly even fatal. By the nature of the tasks performed and the work environment, the laboratory is stressful. Physicians, "stat" laboratory orders, the need for accuracy, lack of communication, errors, and overwork seem to cause clinical laboratory scientists the most stress. Overall, demand for immediate results was found to be the most frequent source of stress, and lack of communication the most intense stressor.[11]

PHYSICAL REQUIREMENTS

Clinical laboratory scientists must have good eyesight or good corrected vision. Good physical health is important. Standing and walking may be required. Manual dexterity and normal color vision are highly desirable.[13, 21]

PROS AND CONS OF THE PROFESSION

Why would these panelists recommend clinical laboratory science as a profession? The future, in spite of current turmoil, represents a true challenge. Laboratory professionals will be given more responsibility for patient management and the health care work environment. The field is rewarding. Clinical laboratory scientists receive a good educational background and learn to solve problems. Their efforts can make a difference to patient welfare. There is still a lot of personal satisfaction.[10]

The profession is exciting despite the gloomy outlook purveyed in a 1985 survey reported in *Medical Laboratory Observer* (MLO).[10] Panelists cite stagnant salaries, dwindling career opportunities, low prestige, nonexistent mobility, and shrinking benefits and budgets as prime reasons for their discontent. Others, however, welcomed what they perceive as an exciting period of growth and change for the laboratory.[10]

Why would MLO panelists not recommend clinical laboratory science as a profession? A majority of the panel would counsel young people to look elsewhere for a career. Is this gloomy outlook valid for today? I think not. The first reason the panelists gave for not recommending CLS as a profession was that jobs were tight in 1985 when this survey took place; in 1990 CLS positions are plentiful. Another reason they would not recommend clinical laboratory science as a profession was low pay. Again, improvements have been made in this area in the past 5 years. Limited room for advancement was the third most frequent reason. The lack of career mobility is a major grievance for clinical laboratory scientists, but it has not prevented many from making frequent job changes.[10]

Given these pros and cons, did the panel recommend a laboratory career to newcomers? Forty-six percent said yes. Of this group, 20% had reservations, and 27% commented that they found their careers satisfying and rewarding. Nineteen percent believe that clinical laboratory science professions are exciting, with fast-breaking developments.

The major reasons clinical laboratory scientists make job changes are to work in another clinical laboratory, to get out of the field altogether, to work for a diagnostic manufacturer, or to work in a nonclinical laboratory. Surveys of clinical laboratory scientists who have left the profession indicate they leave for a variety of reasons. They cite lack of promotion opportunities, inability to fully apply their skills and abilities, routineness of their jobs, and lack of supervisor support. Some believe they are overtrained for their current responsibilities. Many clinical laboratory scientists leave their jobs for various personal reasons; for example, some relocate when spouses are transferred, others put their lab careers on hold for full-time parenthood, and others become full-time students. These facts, along with the career opportunities described in Chapter 11, indicate that career mobility is available with hard work and perhaps advanced education.[9, 10, 19]

Are you still unsure about the profession? Remember your working life in a career may be over 40 years. Take the time to be sure the CLS profession is for you. Volunteer to assist in the laboratory, or perhaps find a position as a phlebotomist or laboratory aide. Familiarize yourself with various clinical laboratory settings and career opportunities in CLS. Collect figures on salaries and benefits. Be realistic about financial rewards. Visit a few laboratories, talk to a number of laboratory professionals, and visit several accredited clinical laboratory science programs. The time you spend investigating the profession is well invested.

The laboratory is a dynamic field with much opportunity for those prepared to grow and change with it. What you get out of the profession is up to you, not the profession. The opportunities in CLS, as detailed in Chapter 11, and my own experiences have convinced me to continue to recommend this profession to students interested in a health profession.

One MLO panelist noted, "Medical technology has a multitude of rewards and some drawbacks, but it is primarily what the individual makes of it."[10]

CHALLENGES OF THE 1990S

In the decade ahead, significant and corresponding challenges are in store for the clinical laboratory and those professionals who work there.[17] Many of these issues are interrelated. The key issues are:

1. Declining student enrollment, CLS program closure.
2. Job satisfaction, turnover of laboratory professionals.
3. Manpower shortage.
4. Prestige, image of the profession.
5. Increasing state and federal regulations.
6. New technology.

Declining Student Enrollment, CLS Program Closure

Analysis of 1988 Committee on Allied Health Education and Accreditation (CAHEA) statistics revealed a decline in the number of graduates of medical technology (CLS) programs of 37% between 1983 and 1988. CAHEA-accredited programs could have accommodated many more students if all available student positions had been filled. The average 1988–1989 mean enrollment was 69% of program capacity.[7]

In 1982, 639 programs offered training in CLS. In 1987–1988 there were only 464 medical technology programs, with an enrollment of 5,706 students and 3,432 graduates. In 1988–1989 there were 436 medical technology programs, with an enrollment of 5,503 students and 3,148 graduates. This represents a 6% decline in accredited programs, a 4% decline in student enrollment, and an 8% decline in the number of graduates in just 1 year. Cytotechnology programs, on the other hand, experienced a 17% increase in student enrollment and 34% increase in graduates during that same year.[8, 14]

In 1988–1989 there were 212 Medical Laboratory Technician associate degree programs, 47 accredited Cytotechnologist programs, 46 Medical Laboratory Technician certificate programs, and 34 Specialist in Blood Bank Technology programs.[8]

No matter how one reviews the data, the clinical laboratory is experiencing a serious shortage of adequately trained personnel. Compelled by new federal reimbursement procedures beginning in 1983, laboratories began to cut costs and medical technologist training programs began to close. In 7 years, from 1982 to 1989, almost 200 accredited medical technology programs closed and enrollment in the remaining programs declined substantially, from almost 9,000 students to around 5,700. There just are not enough students presently graduating from accredited programs to meet the current manpower demands.[17]

Why the decline in student enrollment? For a variety of reasons, the field has not been as attractive a career opportunity as it once was. First, consider demographics: birth rates declined in the 1960s and 1970s, so there are fewer high school graduates from which to draw. In addition, high school and college career counselors have not encouraged students to pursue careers in the allied health sciences. Also, fewer students are going into science.[14, 17]

A growing dissatisfaction on the part of clinical laboratory scientists regarding their profession has diminished recruitment by existing laboratory professionals, contributing to the decline in interest on the part of students. In the past, most clinical laboratory scientists were women. Now more attractive career opportunities for women exist in other areas of medicine and in unrelated fields such as business. As a result, fewer women are choosing CLS as a career.[17]

There is no doubt that CLS was witnessing a shocking decline in graduates even before acquired immune deficiency syndrome (AIDS) emerged as a threat to health care workers. Medically unfounded fears on the part of the general public regarding transmission of human immunodeficiency virus (HIV) no doubt also have caused many prospective clinical laboratorians to consider other career choices.[17]

Studies indicated that, as of April 1988, 15 health care workers worldwide have developed antibodies to HIV after exposure to blood or body fluids from HIV-infected patients. While these reports indicate that transmission of HIV to health care workers by parenteral or mucous membrane exposure is low, many clinical laboratorians understandably have been concerned about the risk of contracting HIV in the workplace. The approximate risk of developing HIV infection after an accidental needlestick injury involving exposure to blood from an AIDS patient is less than 1%.[1]

When asked how the AIDS epidemic has affected their laboratory careers, a survey by Albrecht and Miller[1] showed that 141 (46%) of the respondents felt the same about their career, 92 (30%) felt slightly to moderately dissatisfied with their career, 45 (14%) would never have chosen a laboratory career or are considering changing their career, and 23 (7%) respondents indicated that they felt great satisfaction in providing care to AIDS patients. Most of the laboratory professionals surveyed are not choosing to change their occupation because of AIDS.[1]

The risk for AIDS must be kept in perspective. Clinical laboratory professionals are educated in the safe management and handling of toxic and infectious substances. Extremely reliable, safe procedures are continually being improved on and introduced into practice. With the advent of HIV, these procedures and precautions have been even more intensified, and the universal precautions mandated by the Occupational Safety and Health Administration (OSHA) have minimized drastically any risk of accidental exposure.[13]

Job Satisfaction, Employee Turnover

Some of the same reasons that students no longer choose a career in CLS also are causing clinical laboratory scientists to leave the profession or to change careers. The attrition rate for the profession is reported as 30% to 50%. It should be noted that these dissatisfiers were the same reasons that panelists in the MLO survey gave for not recommending CLS as a profession.[10, 12]

Adding to the dissatisfaction of many clinical laboratory scientists is the separation of their daily work from its end results. The high volume of tests, automation and computerization, and the delegation of direct patient contact tasks to laboratory assistants all have affected clinical laboratory scientists. They get little direct feedback as to whether their contributions have had any bearing on the care of patients.[18]

Job satisfaction has been described as the degree to which the work role and associated working conditions meet the needs and aspirations of the individual or as the degree to which employees see their work-related needs (motives) being fulfilled by the job and the employer.[20] A work role conducive to job satisfaction has the following characteristics:

1. High pay.
2. Substantial promotional opportunities.
3. Considerate and participative supervision.
4. Opportunities to interact with peers.
5. Varied duties.
6. High degree of control over work methods and work pace.[12, 20]

Those in the field who enjoy their work state these reasons for high job satisfaction:

1. Professional pride.
2. Sense of accomplishment in work that is well done.
3. Challenging, interesting work.
4. Sense of independence, little supervision needed.
5. Sense of being needed, making a contribution.
6. Recognition as a care-giver.
7. Developing new technologies.
8. Diverse employment opportunities after graduation and later in the career.

Manpower Shortage

When we consider declining student enrollments and increased employee turnover, the result is a manpower shortage. There is more demand than supply of qualified laboratory professionals. Even the American Society for Medical Technology (ASMT) recognizes this as a major challenge for the 1990s.

Each year in the United States five billion clinical diagnostic tests are requested. The aging population and the expansion of screening programs for substance abuse, AIDS, and other areas have combined to increase laboratory testing. More automated testing methods have not replaced the need for qualified laboratory personnel.

A survey conducted by the American Society of Clinical Pathologists (ASCP) and the ASCP Board of Registry concluded that the nation's 12,000 medical laboratories had 41,950 vacant positions in the summer of 1990. The largest number of vacancies was found for medical technologists, where there was a total of 23,950 vacant positions at staff, supervisory, and managerial levels. The vacancy rate for staff level cytotechnologists was 27.3%.[4]

The vacancy rate in the field seems to depend on which organization is citing the statistics. The ASCP study cited a 11.6% shortage, with one in nine budgeted technologist positions remaining vacant. At the same time, the American Hospital Association (AHA) reported a 5.8% shortage of technologists who have earned at least a B.S. degree. The ASMT reported a shortage of between 10% and 20%.[3, 4, 14]

Hospital laboratories have a higher vacancy rate than private clinical/reference laboratories do. The highest vacancy rates are in small hospitals, those with fewer than 100 beds. The far west central US region seems to have the highest vacancy rates for clinical laboratory scientists. Interesting enough, the west central region and smaller hospitals have the lowest pay rates in general, which suggests a relationship between wages and vacancies.[3, 4]

The state of California has been particularly hard hit. Overall, newspaper ads in California for clinical laboratory scientists have increased 10-fold between 1984 and 1988. In many areas of California, vacancies remain unfilled for months.[6]

What is the outlook for the future? This manpower shortage is likely to continue over the next 5 years, and become more severe over the next few years. The Bureau of Labor Statistics states that "clinical laboratory personnel will have to increase by 24% by the year 2000 to keep up with an expanding laboratory workload." The supply-and-demand situation for the future is discussed in Chapter 14.[6, 14]

Prestige, Image of the Profession

In our society, occupation and social status are interdependent. Society's values and attitudes influence how we perceive an occupation and how much esteem we accord it. Education, required knowledge, salary, and occupational authority and security all contribute to the way we assess prestige on the job.[16]

Who are clinical laboratory scientists or medical technologists? They are "the nurses who do lab tests," as some patients put it. Even sadder, other allied health personnel and clinicians often are just as uninformed about the profession and practice of laboratory medicine.[2]

How do hospital colleagues view the social status of clinical laboratory scientists? A survey of physical therapists (PTs) and occupational therapists (OTs) by the National Opinion Research Center showed that of 13 allied health professions, medical technologist (clinical laboratory scientist) was ranked eleventh. We might assume that respondents would rate their own groups higher due to bias; however, both groups gave their respective colleagues identical social standing, first for PTs and fourth for OTs. The recognition of lower status by OTs themselves indicates a minimal distortion from personal bias.[16]

Clinical laboratory scientists have been plagued by a double-edged image problem. They either have no image at all, which makes them invisible members of the health care team, or they have an image that is false and unflattering. This lack of respect and acceptance is one big reason that we have declining student enrollments and high attrition.[2]

What has caused this image problem? First, lack of awareness. Few outside the field know what laboratory professionals do or how much education is required to do it. Even some laboratory directors and managers do not fully appreciate the intelligence and competence of their staff.

Clinical laboratory scientists have little systematic input and control when it comes to the decision making that affects the quality of their work life. Extra stress is one result, hurting performance and the image of clinical laboratory scientists.

Then, too, many laboratory professionals do not know how to put themselves across well; they do not know the attributes of a respected health care professional. Among other things, they need to learn face-to-face and telephone communication skills.

Another problem is the absence of a centralized association structure in the profession. There is no visible unity. Association strength leads to the independence of an occupation and ultimately professionalization. The CLS profession lacks such strength because of the diverse quantity and quality of its certifying agencies and professional organizations, described in Chapter 7.

Hospital administration can also hold clinical laboratory scientists back. It may fail to provide financial support for continuing education and not encourage clinical laboratory scientists to become team players.[15]

A number of strategies can be used to enhance the professional image of laboratory professionals, principally by providing information or services for others, for example, by sponsoring health fairs, community speaking engagements, and career days. Clinical laboratory scientists themselves can help improve their image by presenting a professional image in dress, demeanor, and speech at all times.[2, 15]

Increasing State and Federal Regulations

The federal government has become increasingly involved in regulating the health care industry. It was the result of governmental influence that led to the increased emphasis on cost containment and quality assurance in health care. Personnel and laboratory licensure changes in the last decade have been the result of government influence as well.

The hospital industry's boom years really began with the advent of Medicare in 1965. During this era of openhanded government support, the more you spent the more you received. There were no incentives for effective cost management, must less cost containment.[5]

The federal government, which now spends an estimated $20 billion to $25 billion a year on health care, began to watch its purse strings more closely in the early 1980s. And while many have predicted the new restrictive financial climate of the hospital would mean less laboratory testing, that has not proved to be the case. In fact, clinical laboratories often are asked to do more tests, with stricter controls, fewer people, and less money.[13, 17]

As a nation we spend $360 billion a year on health care, or 10.5% of our gross national product (GNP). In 1950 we spent only about 5% of the GNP on health care. Medicare and Medicaid programs, funded through the federal government, represent 29% of all health care expenditures. Thus the federal government is the largest consumer of health care services. There were predictions that without intervention the Medicare trust fund would have a balance of zero sometime between 1987 and 1993. Congress responded by approving the Tax Equity and Fiscal Responsibility Act (TEFRA) in 1982 and the Social Security Amendments of 1983. This eventually led to the prospective payment system (PPS) using a diagnosis-related groups (DRGs) system.

The PPS and DRG concept was developed at Yale University in 1970. Every disease and procedure is included within the DRGs. A patient is classified on five factors: diagnosis, age, gender, treatment procedure, and status on discharge. The DRGs are placed in a series of Major Diagnostic Categories (MDCs) directly related to the major body organs.

The PPS is the most revolutionary change in the 17-year life of the Medicare program. The method for payment for Medicare patients has changed from a cost-based retrospective reimbursement system to a prospective rate-setting system based on DRGs. The new system provides incentives to hospitals to manage their operations more cost effectively. The basic element of the PPS program is that the government now pays on a case basis rather than an individual fee-for-service basis. What this means is that when a patient is admitted to the hospital with a particular diagnosis, such as a myocardial infarction, the amount the federal government will pay is already established before the patient receives any treatment. For example, the government already may have established that it will reimburse $4,000 on this DRG, which includes 4 hospital days. If it requires $5,000 and 5 hospital days to treat the disease, the hospital loses money; if it takes $3,000 and 3 hospital days, the hospital makes a profit.

What effect does this have on the laboratory? Prior to 1983, laboratories were reimbursed on a per test basis. Consider a diabetic patient, who generally has high blood glucose levels. If such a Medicare patient required 10 glucose tests as an inpatient, 10 glucose tests were ordered, and the laboratory was reimbursed for 10 glucose tests. Now the reimbursement would be the same whether five glucose tests were ordered or thirty. Some were concerned that the quality of health care would decline. What we have noted is that physician laboratory test ordering patterns have changed. For example, not all patients admitted to the hospital need be tested for syphilis, yet in the past this test commonly was ordered on hospital admission. Laboratories, rather than being "money makers" for the hospital, are now cost centers.

Prior to 1989 the federal government regulated some 12,000 of the estimated 100,000 clinical laboratories in the United States. These regulations helped to ensure that consumers were receiving quality laboratory testing. To assure quality, the government developed regulations requiring quality assurance programs and set minimum personnel standards. However, most physicians' office laboratories remained unregulated. One trend of the 1990s has been a movement of hospital care to alternative delivery systems. There has been a dramatic growth in laboratory sites other than hospital laboratories. In fact, laboratory testing is a $20 billion a year business, with $5 billion of that performed in physicians' office laboratories. Congress in 1988 passed the Clinical Laboratory Improvement Amendments of 1988 (CLIA 88) to regulate these office laboratories. Changes in licensure are discussed in Chapter 6.[17]

New Technology

The impact of biotechnology on the clinical laboratory has been profound. On the one hand, the new technology has added to the increased costs associated with laboratory testing; on the other, these new technologies provide options for diagnosis and treatments that once were considered impossible.

While many people believed that the advent of new technology and DRGs would put clinical laboratory scientists out of jobs, this has not been the case. Automation has changed the nature of the type of work that clinical laboratory scientists perform. Many of the routine tasks associated with paperwork, measuring of reagents, and testing now are performed through the use of automation and robotics. These new technologies require a more sophisticated and more highly trained practitioner to perform the test, analyze the data, and assess the results. Automation has allowed the clinical laboratory to perform more tests with fewer personnel.[17]

REFERENCES

1. Albrecht CM, Miller LE: Impact of AIDS on clinical laboratorians. *Lab Med* 1989; 20:187.
2. Butera RJ, Southerland AR: PR strategies to boost the image of medical technology. *Med Lab Observ* 1985; 17:74.
3. Castleberry BM, Kuby AM, Nielsen L: Wages and vacancy survey of medical laboratory positions in 1990: Part I. *Lab Med* 1991; 22:179.

4. Castleberry BM, Kuby AM, Nielsen L: Wages and vacancy survey of medical laboratory positions in 1990: Part II. *Lab Med* 1991; 22:253.
5. Day CM: Health care macro trends and the laboratory's future. *Med Lab Observ* 1986; 18:45.
6. De Pinna MA: Emerging . . . emerging . . . emerging . . . emerged: The laboratory professional shortage. Employment Opportunities for Medical Technologists and Clinical Laboratory Specialists, California Association for Medical Laboratory Technology 1990; 1:1.
7. Gupta GC: *Allied Health Education Directory 1990*, ed 18. Chicago, 1990, American Medical Association.
8. Gupta GC: Allied Health Education Fact Sheet. Chicago, AMA/Division of Allied Health Education and Accreditation, April 1990.
9. Hajek AM, Blumberg P: Factors contributing to professional attrition of medical technologists. *Lab Med* 1982; 13:488.
10. Hallam K: Part II: Mixed signals for the future. *Med Lab Observ* 1985; 17:37.
11. Jeter J, Chauvin JC, Bailey J: Helping supervisors to reduce stress for clinical laboratory scientists. *Clin Lab Sci* 1989; 2:362.
12. Karni KR, Feickert JD: Occupational reinforcers for medical technologists in clinical laboratories—1966 and 1987. *Clin Lab Sci* 1989; 2:355.
13. Karni K, Oliver JS: *Opportunities in Medical Technology Careers*. Lincolnwood, Ill, 1990, NTC Publishing Group.
14. Knapp J: Help wanted in the laboratory. Collegiate Employment Institute Newsletter, Michigan State University 1990; 4:1.
15. Martin BG, Reyes EL: The low image of MTs as professionals: Reasons and solutions. *Med Lab Observ* 1988; 20:30.
16. Parker HJ, Fong C: The prestige of MTs in health care circles. *Med Lab Observ* 1985; 17:53.
17. Podell LB, Zablotney S: The clinical laboratory in the decade ahead: Challenges and opportunities for ASMT. *Clin Lab Sci* 1990; 3:10.
18. Rosland FA: A study of job enrichment preferences among medical technologists. *J Med Technol* 1985; 2:127.
19. Sharon DJ, Turner BH: Influence of personality on specialty choice and occupational tenure in medical technology. *J Med Technol* 1986; 3:535.
20. Spencer CT, Halinski RS: Statistical analysis of job level and job satisfaction of clinical laboratory scientists. *Clin Lab Sci* 1990; 3:189.
21. US Department of Labor: *Occupational outlook handbook*. Scottsdale, Ariz, 1988, Associated Book Publishers.

Chapter 5

Laboratory Staff, Salaries, and Advancement

LABORATORY STAFF

Pathologist

Roles and Functions.—The director of the medical laboratory usually is a clinical pathologist, a physician trained in laboratory methods to aid in the diagnosis and treatment of disease by scientifically testing the patient's blood, tissues, body fluids, and excretions. The pathologist reports and interprets these findings to the family physician or other attending specialists.

Pathologists may specialize in anatomic or clinical pathology or both, as described in Chapter 3. Anatomic pathologists concentrate on abnormal morphology; clinical pathologists obtain and interpret laboratory data as needed for diagnosis and patient care.[14]

Pathology specialties, concerned with every category of disease, include blood banking/transfusion medicine, cardiovascular pathology, cytopathology, dermatopathology, forensic pathology, hematopathology, immunopathology, and renal pathology.[14]

Surveys show that medical students who choose pathology as a career are often drawn by the opportunities it offers to combine patient care, teaching, and research. Pathology has a special appeal to those who enjoy solving disease-related problems, using technology based in such fundamental biologic sciences as biophysics, molecular biology, and molecular genetics.

Pathologists participate in day-to-day care of hospital patients by providing and interpreting laboratory information to help solve diagnostic problems and to monitor the effects of therapy. The delicate management of patients with acute leukemia, acquired immune deficiency syndrome (AIDS), cancer, and other complex disorders frequently depends in part on the pathologist's medical knowledge and scientific skill.[14]

The clinical pathologist is responsible for clinical laboratory departments such as hematology, chemistry, microbiology, and the blood bank. These activities involve the pathologist in direct patient care as a consultant.

In the clinical hematology laboratory, for example, pathologists review abnormal blood smears. They may also review stained smears and histologic sections of bone marrow to help diagnose diseases such as leukemia.[14]

Pathologists also are concerned with new test development, increasing physician awareness of clinical relevance of test results and appropriate use of laboratory-generated information.

Pathologists may teach medical students, residents in pathology and the clinical services, and graduate students in basic science departments. They may also instruct students in clinical laboratory science and nursing educational programs.

The pathologist-investigator seeks new understanding of the basic nature of disease as a first step toward devising better ways to identify, control, and prevent it. The pathologist in research uses the sophisticated technologies of modern molecular biology and biochemistry.[14]

Education and Certification.—To become a pathologist first requires one to become a physician (M.D. or D.O.). This includes 4 years of college, followed by 4 years of medical school. Medical school graduates currently need 4 to 5 years of residency training after medical school to prepare for a career in pathology. Most pathology residents train in both anatomic (AP) and clinical pathology (CP), although it is possible to train in just one branch. The fifth year of training is needed for combined anatomic and clinical pathology certification. Successful completion of a residency training program and passage of an objective written and practical examination fulfill the American Board of Pathology requirements for certification in AP or CP. After successfully passing the AP or CP examination, a pathologist may become certified in the subspecialties of forensic pathology, medical microbiology, chemical pathology, hematology, blood banking/transfusion medicine, or immunopathology. This generally requires a 1- or 2-year training program or work experience.[14]

Career Options.—Although 75% of pathologists practice in community hospitals, other options include medical schools, independent laboratories, forensics, and the military.

An informative booklet, *Pathology as a Career in Medicine*, is available from the Intersociety Committee on Pathology Information, 4733 Bethesda Ave., Bethesda, MD 20814.

Clinical Sciences Specialists

Roles and Functions.—The clinical chemist and the clinical microbiologist play an increasingly important role in directing laboratory activities. Clinical sciences specialists are generally employed in larger hospital laboratories, research laboratories, or medical centers.[9, 13]

Their job responsibilities are varied. They may be responsible for the management and supervision of a particular laboratory department in their specialty area. They may have patient testing and teaching responsibilities. It is likely they also will be doing research, including new test development, test and method evaluation, or exploring research questions related to the discipline.[13]

Education and Certification.—Specialists usually have a Ph.D. degree in a scientific discipline such as immunology, microbiology, or clinical chemistry. A bachelor's degree in the discipline is the first requirement, plus a master's degree in the discipline. Once accepted into a doctoral program, a student will complete a minimum of 66 hours beyond the master's degree. This usually takes 3 to 5 years and includes a research project.

Certification examinations in Clinical Chemistry are available through the National Registry in Clinical Chemistry and the American Board of Clinical Chemistry. Certification examinations in Clinical Microbiology are available through the National Registry of Microbiologists. Clinical Chemists and Clinical Microbiologists also may qualify for the categorical and/or specialist examinations offered through the American Society of Clinical Pathologists (ASCP) or the National Certification Agency for Medical Laboratory Personnel (NCA). Educational and work experience usually are required to qualify for these examinations. More information on these certification examinations is presented in Chapter 7.[13]

Career Options.—Clinical sciences specialists are primarily employed in private or hospital laboratories; however, private industry, governmental agencies, and educational institutions also offer job opportunities.[13]

Clinical Laboratory Scientist or Medical Technologist

Roles and Functions.—Clinical laboratory scientists (CLS), or medical technologists (MT), have several responsibilities in the clinical laboratory. They may perform supervisory or teaching roles or perform tests to detect diseases. Some clinical laboratory scientists are involved in research.[2, 12]

Laboratory tests play an important role in the detection, diagnosis, and treatment of many diseases. Clinical laboratory scientists perform simple to complex chemical, biologic, hematologic, immunologic, microscopic, and bacteriologic tests. These analyses may require a complex network of steps and variables. Analyses might require operation of complex electronic equipment, computers, and precision instruments costing millions of dollars. Clinical laboratory scientists are trained to understand the scientific basis behind the tests performed in the laboratory.[2, 12]

Clinical laboratory scientists are able to recognize a problem, identify the cause (technical, instrumental, or physiologic), synthesize alternatives, and determine solutions. They are able to confirm and verify results through an in-depth knowledge of techniques, principles, and instruments. They correlate and interpret data based on knowledge of physiologic conditions affecting test results, establish and monitor quality assurance programs, and establish and monitor safety protocols. Clinical laboratory scientists assume responsibility for and are held accountable for accurate results.[4, 7, 11, 18]

Laboratory tests may include chemical tests to determine blood cholesterol levels or microscopic examinations of blood and other substances to detect the presence of diseases such as leukemia. Clinical laboratory scientists microscopically examine other body constituents; make cultures of body fluid or tissue samples to determine the presence of bacteria, fungi, parasites, or other microorganisms; and analyze samples for chemical content or reaction. They type and cross-match blood for transfusions. Technologists may work in all general areas or may specialize in certain areas. Clinical laboratory scientists work independently under general supervision.[4, 18]

Education and Certification.—A clinical laboratory scientist generally has a baccalaureate degree in medical technology, clinical laboratory sciences, or one of the sciences *and* one of the following[16]:

1. Completion of an MT or CLS program accredited by the Committee on Allied Health Education and Accreditation (CAHEA).
2. MLT (ASCP) certification and 3 years of experience.
3. Five years of work-related experience.

See Chapter 9 for additional information.

Certification examinations are available through a number of national organizations, including the Board of Registry of the ASCP and the NCA. See Chapter 7 for additional information.

Career Options.—See Chapter 11.

Clinical Laboratory Technician or Medical Laboratory Technician

Roles and Functions.—A clinical laboratory technician (CLT) performs general tests in all laboratory areas. Working under direct supervision of a clinical laboratory scientist or pathologist, the CLT searches for clues to the absence, presence, extent, and causes of diseases.[10]

The CLT is able to collect, process, and store blood and other patient specimens, perform all of the repetitive (high volume) tests in a modern clinical laboratory, identify direct causes (technical or instrumental) of problems and make corrections using preset strategies, use and monitor quality control procedures, and follow appropriate safety precautions.[2, 4, 16]

CLTs perform all of the routine tests in an up-to-date medical laboratory and can discriminate between closely similar items and correction of errors by use of preset strategies. The technician has knowledge of specific techniques and instruments and is able to recognize factors that directly affect procedures and results. The technician also monitors quality control programs within predetermined parameters.

Some major basic differences between the CLT and CLS in each scientific area are recognized by the Competence Assurance System, American Society for Medical Technology (Table 5-1).

Education and Certification.—The CLT generally has an associate degree in clinical laboratory science from an accredited college or university or a certificate of completion from an approved MLT-C program. Programs include supervised clinical experiences. See Chapter 9 for additional information.

Certification examinations are available through the NCA and ASCP. See Chapter 7 for further information.

Career Options.—Clinical laboratory technicians generally are employed in hospital laboratories, health maintenance organizations (HMOs), clinics, and physician office laboratories.

TABLE 5–1.
Comparison of Roles/Functions of CLT and CLS

Scientific/Technical Area	CLT	CLS
Hematology	Identify normal peripheral blood; recognize presence of abnormal cells	Identify both normal and abnormal blood cells
Microbiology	Identify common organisms	Identify common, unusual, or frequently found organisms
Chemistry	Most chemical testing and routine instrumentation	Routine testing and instrumentation plus knowledge of pathophysiology and operation of sophisticated instrumentation
Immunohematology	Routine testing and antibody identification	Routine testing and identification of difficult, atypical antibodies; investigation of transfusion reactions
Immunology	Serology	Serology and advanced techniques; basic research techniques

Phlebotomist

Roles and Functions.—Phlebotomists draw blood from patients for the purposes of analysis. Most blood is obtained by venipuncture (i.e., from a vein), usually in the arm or from fingers or babies' heels. In the past, most phlebotomists were trained on the job. Now there are approved programs in phlebotomy, offered through a hospital, community college, or university. The programs may vary in length from 6 weeks to 1 year.[9]

Phlebotomists also may have responsibilities in specimen handling and processing. They are responsible for labeling the specimen. They may log the specimen into the computer, centrifuge the blood, prepare aliquots of the specimen according to the specifications of the test to be performed, and deliver the blood specimen to the correct laboratory department.[4]

Education and Certification.—A high school diploma or equivalent is required. Coursework in a phlebotomy program includes medical terminology, ethics, basic anatomy and physiology, specimen collection, anticoagulants, specimen handling, safety, infection control, and interpersonal relations.[9]

Certification examinations for phlebotomists are available through the ASCP and the NCA. Requirements to sit for this examination include high school graduation *and* one of the following:

1. Completion of a National Accrediting Agency for Clinical Laboratory Sciences (NAACLS)–approved phlebotomy program within the last 5 years.
2. Completion of a formal, structured phlebotomy program including at least 120 hours of training.
3. Completion of 1 year of full-time acceptable experience as a phlebotomist.

The certification title for those passing the ASCP examination is PBT(ASCP), Phlebotomy Technician (PBT[ASCP]), and for those passing the NCA examination is Clinical Laboratory

Phlebotomist (CLPlb). A certification examination also is available through the American Society of Phlebotomy Technicians.

Career Options.—Most phlebotomists are employed in hospitals, but other medical facilities, such as physician office laboratories and HMOs, also employ phlebotomists. The current salary for a trained phlebotomist generally is $7.00 to $8.00 per hour. There are opportunities for advancement within the health care setting, particularly with advanced education.

Specialist in Blood Bank Technology

Roles and Functions.—Each year millions of blood transfusions are performed, saving thousands of lives. The specialist in blood bank technology (SBB) contributes in large part to saving these lives. Specialists in blood bank technology perform both routine and specialized tests in blood bank (immunohematology).[7, 13]

Specialists in blood bank technology demonstrate a superior level of technical proficiency and problem-solving ability in such areas as (1) testing for blood group antigens, donor-recipient compatibility, and antibody identification; (2) investigating abnormalities such as hemolytic diseases of the newborn, hemolytic anemias, and adverse responses to transfusion; (3) supporting physicians in transfusion therapy in patients with coagulopathies or candidates for homologous organ transplant; and (4) blood collection and processing, including selecting donors, drawing and typing blood, and performing pretransfusion tests to ensure patient safety. Supervision, management, and/or teaching comprise a considerable part of the responsibilities of the SBB.[7]

Education and Certification.—To be certified an SBB, applicants must (1) be certified in medical technology and possess a baccalaureate degree from a regionally accredited college or university, or (2) if not certified in medical technology, must possess a baccalaureate degree with a major in any of the biologic or physical sciences *and* must complete a 1-year CAHEA-accredited program that covers all phases of blood bank technology.

The educational curriculum is developed to cover all technical areas of the modern blood bank and transfusion services. Programs include extensive practical work experience in a blood bank.[7, 13]

Career Options.—Specialists in blood banking work in many types of facilities, including community blood centers, private hospital blood banks, university-affiliated blood banks, transfusion services, and independent laboratories. Qualified specialists may advance to supervisory or administrative positions or move into teaching or research activities. The criteria for advancement in this field are experience, technical expertise, and completion of advanced education courses.[7]

SALARIES

National salary information resulting from systematic studies over time for occupations within the medical laboratory are not readily available. Most studies on wage information have been limited to "medical technology" positions in hospital and private laboratories. Wages are climbing faster than in many other occupations as the manpower shortage continues. In many areas employers are offering bonuses such as tuition reimbursement, relocation expenses, and sign-on bonuses. Thus the most current information could be outdated before this text goes to print!

An article by Bell and Goldsmith[1] in 1990 gave annual salary ranges for laboratory staff in the Washington, D.C., area:

Laboratory supervisor:	$33,634–$51,334.
Technical/Research Specialist:	$30,181–$47,382.
Senior Medical Technologist:	$27,560–$42,702.
Medical Technologist:	$24,003–$35,984.

One 1989 salary survey reported in *Modern Healthcare* reported starting salaries for medical technologists ranged from $22,380 to $27,892 per year. Starting salaries for medical laboratory technicians ranged from $18,324 to $24,627. By contrast, starting salaries for registered nurses ranged from $22,416 to $32,160, according to American Nurses' Association figures.[5, 15]

Wages paid to clinical laboratory scientists and clinical laboratory technicians varied by location. Of 18 cities surveyed in 1989, the lowest average technologist wage was $10.79/hr in Tampa, Florida; the highest was $16.58/hr in Los Angeles. For technicians, the lowest wage was $8.81/hr in Tampa, compared with $11.84/hr in New York. It is interesting that the highest wages are found in areas that require personnel licensure.[5]

Two more thorough surveys of laboratory workers' wages were sponsored by ASCP and the ASCP Board of Registry in 1990. The pay rates in the ASCP study were conservative measures, reflecting daytime pay rates, and do not take into account differential or shift pay, bonuses, or benefits[3]:

Medical Technologist Staff	$22,069–$30,659
Medical Technologist Supervisor	$26,582–$36,067
Medical Technologist Manager	$31,658–$43,389
Cytotechnologist Staff	$23,566–$33,010
Cytotechnologist Supervisor	$28,538–$40,269
Histologic Technician Staff	$17,992–$24,794
Histologic Technologist Staff	$25,501–$35,277
Histologic Technician Supervisor	$17,680–$24,128
Medical Laboratory Technician	$12,667–$17,888
Phlebotomist	$20,800–$29,411

The second was sponsored by the American Medical Association Division of Allied Health Education and Accreditation (CAHEA). Respondents were program directors of CAHEA-accredited allied health programs. The following 1990 mean entry level salaries were reported[7]:

Medical Laboratory Technician (associate degree)	$18,460
Medical Laboratory Technician (certificate)	$16,861
Medical Technologist Staff	$24,252
Cytotechnologist Staff	

Histologic Technician Staff $19,339
Specialist in Blood Bank Technology $28,867

In general, hospitals with fewer than 100 beds pay the least, and hospitals with more than 500 beds pay the most for staff positions. Large city and suburban hospitals pay higher median salaries than any other employer category surveyed. In general, the pay rates are highest in the far west and northeast regions and lowest in the west central regions.[3]

Progression from a medical technologist (an entry level staff position) through top supervisor represents an increase in constant dollars of $13,998, or 63%. The salary for the top manager could mean an increase of $21,320, or 97%, over a professional career.[3]

Comparing the 1990 ASCP study, the 1989 *Modern Healthcare* report, and the 1990 Bell article, wages for medical technologists have climbed about $2,000 a year over the past few years, an annual increase of approximately 10%.

ADVANCEMENT: THE CAREER LADDER

A career ladder is a sequence of lateral or vertical steps that link jobs related in the same job family, permitting an employee to build on education and experience in order to move to

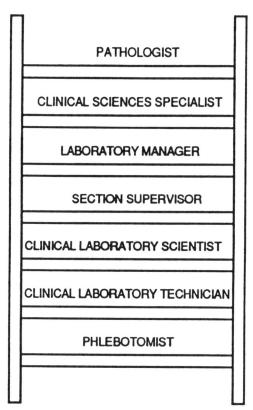

FIG 5–1.
Career ladder in clinical laboratory science.

an advanced position or to a related occupation. Figure 5–1 shows a career ladder in clinical laboratory science. It is possible to be hired as a phlebotomist, and with additional education and experience reach a medical technologist (MT) or clinical laboratory scientist (CLS) position (Fig 5–2).

Medical laboratory technicians may advance to the technologist level by acquiring a baccalaureate degree and passing the MT or CLS certification examination. One route to accomplish this is obtaining a bachelor's degree with the appropriate courses in biologic sciences, chemistry, and mathematics.

To qualify for the MT(ASCP) examination, one of the following is necessary: (1) Complete

---

PHLEBOTOMIST

High school diploma, on-the-job training, or certificate. With completion of a Medical Laboratory Technician program (A. D. or certificate) and passage of a national certification examination:

↓

CLINICAL LABORATORY TECHNICIAN

or

MEDICAL LABORATORY TECHNICIAN

With completion of a Medical Technology program (B. S.) or with completion of a B. S. degree and work experience and passage of a national certification examination.

↓

CLINICAL LABORATORY SCIENTIST

or

MEDICAL TECHNOLOGIST

FIG 5–2.
Career progression from phlebotomist to medical technologist.

an accredited university-based or hospital-based clinical laboratory science program. (2) Qualify through work experience. After passing the MLT examination, one also can qualify with a B.S. degree (including the appropriate sciences) *and* 3 years of full-time acceptable clinical laboratory experience, at least 2 years of which are under the supervision of a pathologist or an appropriately certified medical scientist and a certified medical technologist. One may qualify to take the categorical or the generalist examination, depending on work experience.

Similar educational requirements also will qualify one to take the CLS generalist examination offered by NCA. To qualify to take the CLS generalist examination through work experience, one must have a B.S. degree *and* 2 years of acceptable work experience. It is also possible to qualify to take the NCA examination without a B.S. degree. Sixty semester hours of college credit are required, and 4 years of full-time work experience.

Some university-based CLS programs have articulation agreements with local community

---

MEDICAL TECHNOLOGIST

Automatic upgrade with 3 years of continuous service, better that standard performance:

↓

SENIOR MEDICAL TECHNOLOGIST

With 4 years of experience:

↓

TECHNICAL SPECIALIST

or

TECHNICAL / RESEARCH SPECIALIST

With 6 years of experience:

↓

SUPERVISOR

FIG 5–3.
Career progression from medical technologist to supervisor.

colleges that offer the 2-year CLT program. When students transfer into the university with a CLS program, their transcripts are evaluated and they are given credit for college courses taken, including most CLT or MLT courses. Many of these university-based programs allow CLTs to "test out" of certain courses and shorten their clinical rotation schedules as well. Clinical laboratory technicians may be disappointed in the number of credits transferred into a college or university without a CLS program; most CLT and MLT credits are not transferrable.

The significant obstacle with career ladders is the lack of advancement opportunities for the CLS or MT, particularly for those who do not pursue management, educational roles, or additional education. Some organizations have addressed this issue. Most career ladders for the laboratory are specific to the organization. For example, the career ladder developed by the Washington (D.C.) Hospital Center provides incentives and goals for all levels of laboratory employees. Salary upgrades occur at each rung of the ladder[1, 6] (Fig 5–3).

At Midland Memorial Hospital in Texas, the career steps progress from Graduate Technologist to Medical Technologist I, through Medical Technologist II, Senior Technologist, Section Supervisor, Chief Technologist, and finally, Laboratory Manager. Promotion is based on work experience, continuing education, and good recommendations from the supervisor.[17]

With additional education, one may become a Clinical Sciences Specialist, usually on completion of a doctorate (Ph.D.) in a specialty such as clinical chemistry, microbiology, or immunology. Clinical laboratory scientists may complete medical school and a pathology residency to become pathologists (see Fig 5–1).

The career ladder is a powerful recruitment tool, and can prevent dissatisfaction among laboratory personnel. It offers increasing salaries and a structured career advancement program.[8, 17]

REFERENCES

1. Bell M, Goldsmith G: Some easy solutions work well to ease lab staff shortages. *Advance* Dec 11, 1989; 1:5.
2. *Careers in Medical Laboratory Technology.* Chicago, American Society of Clinical Pathologists, 1989.
3. Castleberry BM, Kuby AM, Nielsen L: Wages and vacancy survey of medical laboratory positions in 1990: Part I. *Lab Med* 1991; 22:179.
4. Clinical laboratory science levels of practice. Competence Assurance System. Washington, D.C., American Society for Medical Technology, 1989.
5. Fitzgibbon RJ: A new look at lab salaries. *Med Lab Observ* 1990; 22:9.
6. Greer KA: Career ladders motivate MTs to climb toward stars. *Advance* 1990; 2:16.
7. Gupta GC: *Allied Health Education Directory 1991*, ed 19. Chicago, 1991, American Medical Association.
8. Hendrix BB: The laboratory career advancement program. *Lab Med* 1989; 20:840.
9. Karni K, Oliver JS: *Opportunities in Medical Technology Careers.* Lincolnwood, Ill, 1990, NTC Publishing Group.
10. *Medical Laboratory Technician: A Career for You.* Chicago, American Society of Clinical Pathologists, 1990.
11. *Medical Technology: A Career for You.* Chicago, American Society of Clinical Pathologists, 1990.
12. Miller-Allan PA: *Introduction to the Health Professions.* Monterey, Calif, 1984, Wadsworth.
13. Nassif JZ: *Handbook of Health Careers.* New York, 1980, Human Sciences Press.

14. *Pathology as a Career in Medicine.* Bethesda, Md, Intersociety Committee on Pathology Information.
15. Perry L: Lab worker shortage getting more attention. *Mod Healthcare* 1989; 19:36.
16. *Preparing for a Career in the Medical Laboratory.* Chicago, American Society of Clinical Pathologists, 1988.
17. Stevenson JW: A career ladder for MT growth. *Med Lab Observ* 1989; 21:43.
18. US Department of Labor: *Occupational Outlook Handbook.* Scottsdale, Ariz, 1988, Associated Book Publishers.

Chapter 6

Registration, Certification, and Licensure

The processes of *registration, certification,* and *licensure* refer to requirements to practice a certain profession; *accreditation* refers to the academic status of a school or organization. All of these processes help assure that health professionals are well trained and qualified.

Over the past century or so, our educational system has established a system of accreditation, to ensure that schools meet appropriate standards of quality.[3] Accreditation means that a school's credits can be transferred to other schools and used to meet the qualifications for entering various professions. In the clinical laboratory science profession, graduation from an accredited program permits one to sit for many of the certification examinations for medical technologists or clinical laboratory scientists.

The companion methods of credentialing members of the health occupations in the United States are licensure (practice control) and certification (title control). Each identifies differing objectives, requirements, and methods of administration; together they exert an enormous impact on the health care system. Recent decades have seen an increasing demand for licensure on the part of health occupational groups. This increase reflects a sense that licensure grants recognition to professionals and provides the public with a feeling of protection. These two credentialing processes are completely separate. Licensing is done by state agencies; certification is done by organizations of professionals.

Certification is "a process in which an individual, an institution, or an educational program is evaluated and recognized as meeting certain predetermined standards. Certification is usually made by a nongovernmental agency. The purpose of certification is to assure that the standards met are those necessary for safe and ethical practice of the profession or service."[12] Certification is voluntary and serves the private sector as recognition of competence or excellence in practice.

Registration is a general term. It may mean title control or it may mean simply that a law requires all persons who wish to engage in a given occupation to register with a designated government agency. Registration usually involves listing one's name and address and payment of a fee. Registration is used in situations where the threat to public health, safety, or welfare is minimal. It is relatively easy to become registered.

Although the term "registered medical technologist" is used frequently, the correct term is *"certified medical technologist."* Registration implies only that a person's name is on a register. It is assumed that the person who refers to himself or herself as a registered medical technologist has completed a prescribed course of study and passed a certification examination.

"Licensing is the granting of permission by a competent authority (usually a governmental agency) to an organization or individual to engage in a practice or activity that would otherwise be illegal. Kinds of licensure include issuing of licenses for general hospitals or nursing homes, for health professionals, as physicians, and for the production or distribution of biological products."[12] The state chooses to regulate some health occupations to ensure that public safety, health, and welfare are reasonably protected.

Since the law establishing a licensed occupation usually sets forth the "scope of practice" covered by the act, licensing is often referred to as a "practice act." Personnel licensure and laboratory licensure are presented separately.

Title licensure or title control is another mechanism for credentialing health professionals. This simply means that the title is licensed, but not the practice.[8] For example, consider the clinical laboratory scientist. Title certification means that only those individuals who have completed a prescribed course of study and passed a certification examination may call themselves clinical laboratory scientists. Other individuals can perform laboratory work, but may not refer to themselves as clinical laboratory scientists. Consumers decide whether they prefer someone who is licensed or certified.

PERSONNEL LICENSURE

Licensing has essentially two functions. First, the licensing of occupations and professions is a means of protecting the public. Second, in addition to protecting the public, licensing can serve to protect the license from competition by legally excluding from the practice of an occupation those who have not acquired the appropriate training as defined by members of the occupation.

As a laboratory professional, you will make a decision regarding personnel licensure. Consider the pros and cons.

Pros

Salary Increases.—If a health professional must be licensed to practice in a particular state, supply and demand is affected. Unlicensed personnel may no longer be employed. With fewer individuals available for employment, supply is limited and demand is high. Thus salaries tend to increase to attract licensed professionals.

Improved Respect and Prestige as a Health Care Team Member.—Laboratory personnel tend not to be recognized for their contributions to patient care. In fact, many patients and physicians may be unaware of the education and training required to become a clinical laboratory scientist. Many believe the respect and prestige afforded registered nurses results from state licensure laws.

Improved Quality of Laboratory Testing.—Studies indicate that qualified laboratory personnel produce more accurate laboratory results. Accurate laboratory testing is critical for the care and treatment of patients.[13]

Increased Visibility of Laboratory Personnel to Public.—As states develop personnel licensure, the media and legislators become more aware of the vital role laboratory personnel play in the U.S. health care system.

Defined Area "Content" is Ascribed to the Profession.—Groups seeking to become licensed usually want to define their scope of practice broadly. They want to stake out as much territory as possible to keep other groups from encroaching. For example, both respiratory therapists and clinical laboratory scientists may include within their scope of practice determination of arterial blood gas and electrolyte levels. However, if only respiratory therapists are licensed and these tests are included in their scope of practice, it would be illegal for clinical laboratory scientists to perform these tests in that state.

Cons

Worker Mobility Becomes Limited.—Because one is licensed to practice in one state does not imply that he or she legally can practice in another. Thus frequent moving from state to state can become burdensome to maintain licensure. One response to the problem is state "reciprocity"; that is, one state will recognize the licenses from another state, and vice versa. Reciprocity agreements between states are common for physicians and registered nurses. At this time, there are no reciprocity agreements related to laboratory personnel.

Increased Cost to Personnel to Maintain Licensure.—Health care professionals are responsible for costs associated with licensure. There is a charge for the state certification examination, the initial certification, and for recertification.

Increased Cost to the State to Implement and Maintain Personnel Licensure.—Many state legislators and Bureaus of Licensing and Standards are opposed to further certification of health professionals, because of the high cost to the state to develop valid tests, hire additional personnel, and maintain paperwork when a profession is licensed.

Possible Decline in Health Care Available in Rural and Underdeveloped Areas.—Rural and underdeveloped areas have difficulty in attracting and keeping qualified laboratory personnel. If state licensure exists, there is even greater difficulty. Access to health care by the poor and minorities may become more limited as a result.

Increased Costs of Laboratory Testing.—It is likely that if qualified higher paid individuals perform laboratory testing, these increased personnel costs will be passed onto the consumer.

Update

The credentialing of professionals in laboratory medicine is of paramount importance. Credentialing may contribute to standards of quality and excellence in the laboratory.[4] Licensure first was offered for laboratory technologists in California in 1938. California medical technologists believe there is a direct relationship between its strict licensing law, difficult examinations, and higher salaries (generally $2,000 to $5,000 higher than in nearby states).[6] About 15 years ago, interest in licensure began to develop in other states.

TABLE 6–1.
State Licensure of Laboratory Personnel*†

All personnel	**No licensure requirements**
California	Alabama
Florida	Alaska
Hawaii	Arkansas
Nevada	Colorado
Tennessee	Idaho
W. Virginia	Indiana
N. Dakota	Iowa
New York City	Kansas
	Louisiana
Director/Supervisor	Massachusetts
Connecticut	Minnesota
	Mississippi
Laboratory/Director	Missouri
Delaware	Montana
Georgia	Nebraska
New Jersey	New Mexico
New York	N. Carolina
Rhode Island	Ohio
	Oklahoma
Laboratory only	S. Carolina
Arizona‡	S. Dakota
Illinois	Texas
Kentucky	Utah
Maine§	Vermont
Maryland	Virginia
Michigan	Washington‖
New Hampshire	
Oregon	
Pennsylvania	
Wisconsin¶	
Wyoming¶	

*Adapted from Daley DM: State licensure for laboratory professionals: A survey. *Lab Med* 1984; 15:556.
†NOTE: The District of Columbia is under federal standards only.
‡Director must qualify.
§Independent laboratories only.
¶Directors must obtain certificates.
‖Personnel standards were included in a Medical Test Site Licensure Bill passed in 1989.

Personnel standards vary widely among the states. In Arizona, only a physician or osteopath may serve as laboratory director. In California, a physician or bioanalyst may direct a multiservice laboratory. A supervisor in Connecticut needs at least a bachelor's degree, but in Oregon supervisors are eligible with 3 years of college with laboratory experience.[7]

Only seven states—California, Hawaii, Tennessee, Nevada, Florida, West Virginia, North Dakota—and New York City have laws requiring licensure of all personnel, with exceptions at the level of technicians. Professionals wishing to be employed in these areas should contact the Department of Health or suitable governmental agency with responsibilities in health personnel licensure. Generally an individual can be granted a temporary license to practice until able to meet state requirements.

States currently without licensure laws include Alaska, Colorado, Idaho, Indiana, Iowa, Louisiana, Missouri, Montana, Nebraska, New Mexico, Ohio, Oklahoma, South Dakota, Vermont, and Virginia.

Other states considering licensure or that have introduced bills to their state legislature include Georgia, Illinois, Kansas, Louisiana, Maine, Massachusetts, Michigan, Mississippi, Missouri, Nebraska, New Mexico, Oregon, Texas, Utah, and Washington. Efforts toward licensure have failed in these states because of lack of consensus within the field as to the value, need, or criteria for licensure.

Personnel standards under federal licensure (for independent and interstate laboratories and those receiving Medicare reimbursement) are specifically defined. State requirements apply to all laboratories, even those not covered by federal programs. State rules frequently impose additional personnel requirements. State laws, in many instances, contain complicated grandfather clauses or alternate requirements for compliance. A grandfather clause commonly included in personnel licensure bills allows currently employed laboratory personnel to be licensed because of their current work experience. However, new professionals must meet licensure requirements, usually including educational requirements and certification.

The scope of state licensure for laboratories varies as much as the conditions for eligibility. Currently 16 states have legislation, and 11 states have regulations regarding laboratories not regulated by the federal government.[9] Some states regulate only independent laboratories, others only hospital laboratories, and still others apply separate standards to the two categories. Some states do not license at all, but regulate hospital laboratories through hospital laws. And in a few states, strict rules are on the books but are not enforced. Table 6–1 shows the status of state laboratory and personnel licensure, current as of 1990.

LABORATORY LICENSURE

Many laboratories are licensed because of federal mandates. Hospital laboratories, independent reference laboratories, and laboratories participating in interstate commerce or receiving Medicare funds must meet certain federal guidelines. In addition, many laboratories choose to participate in voluntary accreditation programs; examples include the Joint Commission on Accreditation of Healthcare Organizations (JCAHO), the College of American Pathologists (CAP), and the American Association of Blood Banks (AABB).[10]

Accuracy of laboratory testing has been the focus of much media attention over the past few years. The *Wall Street Journal, USA Today,* and national television news reports have raised questions about the quality of laboratory testing. The federal government is imposing external regulatory requirements on independent physician office laboratories.

States that developed laws regulating physician office laboratories 5 or more years ago generally based regulation on the number of physicians in practice together, exempting office laboratories that served fewer than three or four physicians. Things began to change in 1980 when Pennsylvania pioneered a new three-tier system that regulated office laboratories based on the difficulty of the laboratory tests performed on-site.[2]

FEDERAL REGULATION OF LABORATORIES

As a result of challenges to the quality of services provided by laboratories in the United States the federal government passed the Clinical Laboratories Improvement Act of 1967 (CLIA '67).[10] The provisions of this act are applied only to laboratories that were in interstate commerce. Laboratories were prohibited from operating in this sphere unless they had an appropriate license issued by the Department of Health, Education and Welfare (now Department of Health and Human Services, DHHS) or its designee. Of the approximately 12,000 federally regulated clinical laboratories in the United States, 6,700 are located in hospitals.[5] Exempted are clinical laboratories operated by a licensed physician, osteopath, dentist, podiatrist, or group thereof, who perform laboratory tests or procedures solely as an adjunct to treatment in their own patients.

Laboratories are regulated if they receive Medicare reimbursement (Title XVIII of the Social Security Act). Medicare laboratory regulations apply to hospital and independent laboratories that qualify to receive reimbursement for performing laboratory tests for patients in both the Medicare and Medicaid programs.[5]

The federal regulation of physician office laboratories will be implemented as a result of the Budget Reconciliation Act of 1987 (OBRA '87) and The Clinical Laboratory Improvement Amendments of 1988 (CLIA '88). This act mandates that *all* physician office laboratories performing 5,000 or more tests per year meet Medicare performance standards required of independent laboratories. CLIA '88 specifies that its provisions were to go into effect on January 1, 1990, for most laboratories, and on July 1, 1991, for physician office laboratories. This provision means that laboratories performing more than about 20 tests per day are subject to the same quality control, proficiency testing, record-keeping, and personnel standards required of larger, professionally staffed and directed laboratories.[2, 11]

The Health Care Financing Administration (HCFA) has the responsibility for implementing CLIA '88. Some of the greatest objections have been raised to the HCFA proposal for regulating laboratories based on test complexity as required by CLIA '88. CLIA '88 authorizes the DHHS to grant certificates of waiver to physicians' office and other small laboratories that perform only "simple laboratory examinations and procedures which, as determined by the secretary [of DHHS], have an insignificant risk of an erroneous result."[11]

The HCFA proposed rules would divide clinical laboratory tests into three levels on the basis of complexity. The first, or "waiver," level includes tests that do not pose a risk for harm to the patient if performed incorrectly, are unlikely to yield incorrect results, and require simple methods. Physicians' offices and other sites that perform only waiver-level tests would be exempt from regulation.[11] Level 1 tests, according to HCFA, pose some risk if results are incorrect, but that risk is minimized because the testing methods are not complex. Level II tests include all those not in the first two categories.[11]

What effect will these laws have on the profession? Individuals presently employed in physician office laboratories will have to meet federal personnel requirements, if they do not already. Some may choose to leave their position; others may return for more education. It is likely there will be an increasing demand for certified laboratory workers with a B.S. or A.D. degree.

REFERENCES

1. American Society for Medical Technologists: State licensure update. Unpublished material, 1989.
2. Baer DM, Skeels MR, Belsey R: Regulation of physicians' office laboratories. *Med Lab Observ* 1988; 20:26.
3. Barrett S: Should "nutritionists" be licensed? The time is right (editorial). *Postgrad Med* 1986; 79:11.
4. Daley DM: State licensure for laboratory professionals: A survey. *Lab Med* 1984; 15:554.
5. Kenney ML, Greenberg DP: Final report on assessment of clinical laboratory regulations. Unpublished material, 1986.
6. Kull DJ: California's model licensure law. *Med Lab Observ* 1978; 10:100.
7. Kull DJ: State licensure laws for laboratorians. *Med Lab Observ* 1980; 12:75.
8. Miller J: Licensure, credentialing, or registration? Drawing the lines pro and con. *Respir Ther* 1982; 12:31.
9. Office of Inspector General: Quality assurance in physician office laboratories. Unpublished material, 1989.
10. Snyder JR, Larsen AL: *Administration and Supervision in Laboratory Medicine*, ed 2. New York, 1989, Harper & Row.
11. Tune L: Initial CLIA '88 rules released: Critics unmoved. *Clin Chem News* 1990; 16:1.
12. Urdang L (ed): *Mosby's Medical & Nursing Dictionary*, St Louis, 1983, CV Mosby.
13. Wobbe R: Institutional licensure versus individual licensure. *Texas Nurs* 1978; 52:10.

Chapter 7

Certification Agencies, Accrediting Agencies, and Professional Organizations

CERTIFICATION AGENCIES

Certification is the process by which a nongovernmental agency or association grants recognition of competence to an individual who has met certain predetermined qualifications as specified by that agency or association.[2]

Board of Registry of the American Society of Clinical Pathologists

History

With the increasing demand for well-trained laboratory workers, the American Society of Clinical Pathologists (ASCP), working with the American Medical Association (AMA) Council on Medical Education, established the Board of Registry in 1928. The first board consisted of six pathologists. Later, the board became a standing committee of the ASCP. With the establishment of this board, all graduates of medical technology training programs were required to pass its examinations. The first certificates were awarded by the board in 1930. Requirements for certification included graduation from high school or a diploma from a school of nursing and 1 year of training.[5, 6, 8, 9]

In 1933, applicants were required to pass an essay examination and an oral and practical (hands-on) examination before certification was granted. In 1934, educational requirements were changed to 2 years of college, including science courses, and 1 year of clinical training (2 + 1). In 1962, 3 years of college and 1 year of clinical experience were required, thus supporting the development of B.S. degree programs.[5, 6, 8, 9]

In 1963 ASCP began providing certification examinations for laboratory technicians. By the end of 1990 the Board of Registry had certified a total of 291,218 individuals, of whom 184,013 were registered as medical technologists.[3, 7]

The ASCP also has provided input on program accreditation. After its organization in 1933, the American Society for Medical Technology (ASMT) joined ASCP in periodic revisions of program accreditation. In 1949 the Board of Schools was formed by the Board of Registry. The Board of Schools then became the accrediting agency for medical laboratory training programs. This board was replaced in 1973 by the National Accrediting Agency for Clinical Laboratory Sciences (NAACLS), which was sponsored by the Board of Registry. NAACLS now has the role of accrediting laboratory training programs.[9]

The Board of Registry of ASCP first offered the examination for cytology technician in 1957. Until 1975, cytotechnology programs were accredited through a joint effort of NAACLS and the ASCP Board of Registry. In 1975, the American Society of Cytology (ASC) was recognized as the organization that would collaborate with the AMA Council on Medical Education in cytology program accreditation matters. Since 1975 this board has assumed the responsibilities formerly handled by NAACLS.

In 1977 the composition of the Board of Registry was changed to include six pathologists, six medical technologists, two lay persons, and a representative from each of the six participating organizations of laboratory specialists. These six participating specialty societies include the American Academy of Microbiology, the American Association of Blood Banks, the American Society of Cytology, the American Society of Hematology, the National Registry in Clinical Chemistry, and the National Society of Histotechnology.[2]

Functions

In conducting certification activities, the Board of Registry:

1. Prepares standards that assure competence of medical laboratory personnel.
2. Develops and administers appropriate examinations for measurement of competence.
3. Develops and uses appropriate scoring procedures.

Certifications Available

The Board of Registry offers 18 examinations for four basic types of certification: general, categorical, specialist, and diplomate. The first includes medical technologists, medical laboratory technicians, and phlebotomists.

Categorical certifications are available in blood banking, chemistry, microbiology, hematology, and immunology. In general, to qualify for these examinations one must have a baccalaureate degree and related work experience. Certification examinations are also available in nuclear medicine technology, histology, and cytology.

Specialist certifications are available in hematology, microbiology, chemistry, and cytotechnology. One may qualify to take these examinations with a bachelor's, master's, or doctoral degree. Those with a bachelor's degree must have MT(ASCP) certification or appropriate categorical certification. There is a minimum work experience requirement that depends on the education of the applicant. For example, a person with a bachelor's degree must have 5 years of experience; with a master's degree, 4 years; and with a doctorate, 2 years of related work experience. Up-to-date requirements may be obtained from the Board of Registry, ASCP.[2]

A Diplomate in Laboratory (DLM (ASCP)) Management is also available to those who satisfy educational and work experience requirements and successfully pass the examination.

National Certification Agency for Medical Laboratory Personnel

In 1977 the American Society for Medical Technology helped establish a new voluntary, nonprofit certification agency, independent of control by any professional association. Prior to that time, the Board of Registry of the ASCP controlled certification of the majority of laboratory personnel. Many laboratory professionals felt the need to establish autonomy apart from pathologists, and initiated the National Certification Agency for Medical Laboratory Personnel (NCA). In 1978 NCA offered its first examinations in two generalist categories: Clinical Laboratory Scientist (CLS) and Clinical Laboratory Technician (CLT).

The NCA now provides examinations in the following categories: Clinical Laboratory Scientist (CLS), Clinical Laboratory Technician (CLT), Clinical Laboratory Phlebotomist (CLPlb), Clinical Laboratory Specialist in Cytogenetics (CLSp(CG)), Clinical Laboratory Specialist in Hematology (CLSp(H)), Clinical Laboratory Director (CLDir), and Clinical Laboratory Supervisor (CLSup). To date, more than 65,000 practitioners have successfully completed the NCA certification process. NCA certification examinations are administered nationally on the last Saturday in January and July each year.

The NCA, unlike the Board of Registry, requires recertification of clinical laboratory personnel. This can be accomplished by reexamination or by documentation of participation in continuing education (4 continuing education units [CEUs] within 4 years).[7]

American Medical Technologists

Founded in 1939 as a nonprofit organization, American Medical Technologists (AMT) provides national certification to four categories of registrants: medical technologists, medical laboratory technicians, medical assistants, and dental assistants. In addition, AMT also serves as an accrediting agency by approving laboratory education programs through the Accrediting Bureau of Health Education Schools (ABHES). AMT certification requirements focus on technical training, educational background, and work experience. Its medical technologist examination can be taken by persons with a baccalaureate degree as well as by AMT technicians with 3 years of experience. Members of AMT tend to be graduates of 1- or 2-year proprietary schools who first qualified for the technician (MLT) category and then, with work experience and passing of the AMT technologist's examination, gained medical technologist (MT) status.[6]

The AMC publishes the *Journal of American Medical Technologists*, which contains scientific papers, Registry (Society) news, legislative and industry reports, book reviews, and editorials.

AMT serves as a professional organization, providing its members with professional publications, continuing education programs, free job assistance services, and concerned representation in state and federal legislative matters. AMT coordinates an ongoing certificate revalidation program. AMT registrants are encouraged to revalidate their certificates every 5 years. This requires 70 hours of continuing education credits within 5 years, completion of 6 semester hours in occupation-related programs, or 100 clock hours of professional growth activities such as article preparation, or successful completion of a revalidation examination.

Since its beginning, AMT has certified more than 27,000 medical technologists. Its active membership is in excess of 21,000 as of 1990.

International Society for Clinical Laboratory Technology

The International Society for Clinical Laboratory Technology (ISCLT) serves as an autonomous certification agency for Medical Technologist (RMT), Laboratory Technician (RLT), and Physician Office Laboratory Technician (POLT). Founded in 1962, ISCLT also serves as a professional society of laboratory personnel, with membership composed primarily of persons trained on the job. By the end of 1988, ISCLT had an active membership list of approximately 19,000.

ISCLT provides certification examinations through its Credentialing Commission. ISCLT examinations are available to persons with laboratory experience and who are sponsored by a supervisor. All individuals certified by the Credentialing Commission are required to revalidate their certification periodically. This revalidation of certification is done annually by the Credentialing Commission, which reviews activities of each certificant and reissues a certificate.

The ISCLT has a different philosophy from that of ASMT and AMT. It believes the working place environment is a valid learning site and that learning takes place in such environments as well as in academic institutions. Therefore, admission to the Credentialing Commission examinations may be attained through alternate routes: academic attainment, specialized training programs, work experience, or a combination of these. As a result of this philosophy, it is one of the few laboratory organizations that recognizes proficiency examinations offered by the DHHS.[7]

As a professional organization, ISCLT sponsors continuing education programs (Continuing Education for Professional Advancement, CEPA) through the Credentialing Commission. ISCLT also publishes a newsletter and a recording service for continuing education activities for its members.[7]

Department of Health and Human Services

To further complicate certification efforts, the federal government began proficiency examinations for support personnel in 1975. The intent of these examinations was to ensure that independent laboratories performing testing for Medicare patients had sufficient "properly qualified personnel." Because of stringent requirements, many independent laboratories and hospital laboratories did not meet Medicare standards. The Department of Health, Education and Welfare (DHEW), now DHHS, tried to upgrade the status of workers employed through examination, rather than downgrade its original standards. More than 50,000 support personnel and cytotechnologists from a variety of laboratory settings have taken the DHEW examination. Approximately half have passed and have been certified by DHEW as clinical laboratory technologists or cytotechnologists.[6]

INDEPENDENT CERTIFICATION AGENCIES

A number of independent registries have been established in laboratory specialties.

National Registry in Clinical Chemistry

The National Registry in Clinical Chemistry (NRCC), established in 1967, is sponsored by eight boards and associations of chemists and pathologists. The Registry grants certification in three categories: Clinical Chemistry Technologist, Clinical Chemist, and Toxicological Chemist. The category of Clinical Chemistry Technologist is designed primarily for applicants with recent bachelor's degrees who regularly perform clinical chemistry determinations. The category of Clinical Chemist exists for college graduates with extensive clinical chemistry experience or advanced education. The category of Toxicological Chemist exists for college graduates with extensive toxicology experience or postgraduate education. All applicants must take an examination designed to test their practical knowledge of the fundamentals of clinical chemistry.[7]

National Registry of Microbiologists (NRM)

See American Society for Microbiology, p. 61.

American Board of Clinical Chemistry

The American Board of Clinical Chemistry (ABCC) was incorporated in 1950 as an independent organization. Its functions include establishing and enhancing standards of competence in clinical and toxicologic chemistry and certifying qualified specialists. Diplomate certification programs are available in Clinical Chemistry and Toxicological Chemistry. Certification is based on a candidate's education, experience, and certification examination score.

The ABCC serves as a sponsoring organization of the National Registry in Clinical Chemistry.

ACCREDITATION AGENCIES

Accreditation of educational programs is a separate and distinct process, different from certification of personnel. There is, however, a relationship between the quality of the educational process and the caliber of the graduate of that process.[7]

Accreditation is a process of external peer review in which a private, nongovernmental agency or association grants public recognition to an *institution* or *specialized program of study* that meets certain established qualifications and educational standards as determined through initial and subsequent periodic evaluations.[5]

The purpose of the accreditation process is to provide a professional judgment of the quality of the educational institution or program and to encourage its continued improvement.

National Accrediting Agency for Clinical Laboratory Sciences

In 1972, representatives of ASMT and ASCP began talks that culminated in the incorporation of the National Accrediting Agency for Clinical Laboratory Sciences (NAACLS) in October 1973. It is an organization independent of any professional organization and to which the authority of program accreditation is delegated. NAACLS is the first agency of its kind to conduct the processes of detailed program review and recommendations on which the Committee on Allied Health Education and Accreditation (CAHEA) accredits programs.[5]

Units within NAACLS include the Review Board, three program review committees, and the executive office staff. The Review Board functions as the governing unit and reviews all programs. The executive office staff serves to facilitate the program accreditation process. Several review committees report to the NAACLS board, each committee having responsibility for reviewing one of the types of laboratory educational programs within the NAACLS purview. NAACLS serves as a review body for programs in Medical Technology, Medical Laboratory Technician (associate degree), Medical Laboratory Technician (certificate), Histotechnology, and Phlebotomy. The NAACLS executive office arranges for accreditation site visits to educational programs, assembles reports to go to the board for action, and forwards recommendations from the board to CAHEA.[7]

The bylaws of NAACLS specify board representation for sponsoring organizations (ASCP, ASMT) and participating organizations (currently, American Society for Microbiology, National Society for Histotechnology). In addition to the representatives from these professional organizations, the membership of the board includes two college educators, two public members, a laboratory director, a technologist laboratory administrator, and two technician practitioners.[5]

In addition to the work associated with conducting program reviews, NAACLS provides workshops and publishes a periodic newsletter, information for students, and occasional monographs addressing issues in accreditation.[5]

Committee on Allied Health Education and Accreditation

The Committee on Allied Health Education and Accreditation (CAHEA), in cooperation with 19 program review committees, accredits educational programs in 26 allied health occupations, including Medical Technologist, Cytotechnologist, Histologic Technician, Specialist in Blood Bank Technology, and Medical Laboratory Technician. The AMA sponsors CAHEA. Allied health professional organizations and medical specialty societies sponsor the review committees. Institutions sponsoring allied health programs participate in CAHEA to meet national accreditation standards to prepare allied health personnel. The 14 members of the CAHEA Board include persons with a broad interest and/or competence in the allied health professions.[5]

Accreditation recommendations are submitted by a review committee, such as NAACLS. These recommendations are reviewed by one of two CAHEA subcommittees, which give special attention to recommendations for placing programs on probation or for withdrawing or withholding accreditation. These committees are designed to ensure that due process and policies and practices have been followed in arriving at the accreditation recommendations.[5]

Once the subcommittee work is completed, CAHEA convenes to award or deny accreditation based on the recommendations resulting from the subcommittee review. A formal notice is then sent to the chief executive officer of the sponsoring institution.[5]

CAHEA, in cooperation with the review committees, is recognized nationally as an "umbrella" accrediting body by the U.S. Department of Education and the Council on Postsecondary Accreditation (COPA) for allied health educational programs. As a national, nongovernmental, nonprofit educational organization, COPA coordinates voluntary accrediting activities for institutions and programs at the postsecondary level.[5]

Accrediting Bureau of Health Education Schools

The American Medical Technologists (AMT) organization has sponsored and supported the growth of the Accrediting Bureau of Health Education Schools (ABHES), which accredits training programs for medical laboratory technicians and medical assistants. This agency was established as an autonomous body, and has focused its mission toward accreditation of private and proprietary allied health education schools. The Board of Commissioners of ABHES is composed of a different mix of representatives than that of the NAACLS Review Board, and makes final decisions regarding accreditation of programs rather than forwarding recommendations to another agency.[7]

PROFESSIONAL ORGANIZATIONS

A professional organization should reflect the voice and beliefs of its members. Its purpose is to increase the recognition and prestige of the group and to represent the profession to the public, to governmental agencies, and to other professional organizations. Every laboratory professional should be encouraged to join a professional organization. Of the number of professional organizations offering membership, the American Society of Medical Technologists (ASMT) probably best reflects the views of clinical laboratory scientists and technicians.

American Society for Medical Technology

The American Society of Clinical Laboratory Technicians was organized in 1933, and held its first national convention that year. The name was changed to the American Society of Medical Technologists in 1936, and again in 1972 to the American Society for Medical Technology (ASMT).[7]

The mission of the ASMT is "to improve the public's health through promotion of efficient and effective use of clinical laboratory services through: (1) achieving optimal utilization of laboratory sciences; (2) serving as laboratory information specialists; (3) ensuring an adequate supply of competent laboratory practitioners; (4) promoting practice at a level commensurate with the expertise of the profession and removing barriers to such practice; and (5) explaining the role of laboratory personnel to consumers, third-party payers, and other health practitioners."[1]

Qualifications for active membership include a baccalaureate/graduate degree in medical technology or related science; or a license and/or certificate from an agency recognized by ASMT in an area of clinical laboratory science; and/or completion of an accredited structured program in CLS.

The membership has grown from about 500 in 1937 to more than 20,000 in 1990. Members include directors, supervisors, managers, educators, consultants, and bench technologists. Eighty-one percent of ASMT members hold a baccalaureate degree, and a significant number hold higher degrees.

The national organization, to which every member belongs, is governed by a board of directors responsible for management of the properties, funds, and activities of the Society. The House of Delegates is the governing body of the Society. Its membership consists of delegates from each constituent society, and the members of the Board of Directors.[1]

ASMT is divided geographically into 10 regions made up of state societies. In addition to belonging to the national organization, ASMT members become members of the regional, state, and local societies. Additional dues may be required to belong to the state and local societies.

The growth of ASMT over the years has accompanied a gradual trend toward professional independence for clinical laboratory scientists. ASMT has earned national recognition because of its promotion of professionalism and for enhancement of the stature of the medical technologist as a member of the health care team. These later accomplishments are perhaps the most significant, noteworthy, and deserving of the appreciation of members and nonmembers alike.

ASMT has become active in government relations over the past few years. Headquartered in Washington, D.C., ASMT maintains continuous liaison with Congress and the federal executive branch. The government affairs staff monitors and influences federal and state laws and regulations that have an impact on its members. Through the ASMT Political Action Committee (PAC), members have a legal, ethical way to pool political campaign contributions to support the election campaigns of U.S. senators and representatives of both parties.

Membership is not based on credentials, and membership categories include associate and student levels in addition to active membership.

A major educational service offered by ASMT is recording the continuing education activities of members and other laboratorians who desire to enroll in the recording program. Titled Professional Acknowledgment for Continuing Education (PACE), this ASMT service reviews and approves hundreds of continuing education opportunities each year in ASMT publications and at local, state, regional, and national meetings. CEUs earned are automatically recorded on the individual's PACE record.[7]

Members of ASMT receive the peer-reviewed, bimonthly, professional journal *Clinical Laboratory Science* and a bimonthly newsletter, *ASMT Today*. Members are able to attend state, regional, and national meetings at reduced rates, and to participate in the Employment Opportunities Program (EOP), which attempts to bring the most qualified job applicants together with the nation's leading clinical laboratory science employers. ASMT also has an expanding library of peer-reviewed publications at substantial discount. Insurance plans and financial programs (MasterCard and Member Loan Program) are additional benefits available to members.

In the early 1980s, a new level of cooperation developed between ASMT and AMT. A jointly sponsored national meeting was held, and a jointly published journal was made available to ASMT and AMT members. This progress has not continued as the 1990s begin. The obvious benefits of the merger include strengthening membership benefits and greater political influence.

American Society of Clinical Pathologists

The American Society of Clinical Pathologists (ASCP) is the professional organization devoted to the interests of those engaged in the medical specialty of pathology. It is a not-for-profit corporation organized exclusively for educational, scientific, and charitable purposes. Several classes of membership are offered. Full membership is granted to pathologists who have been certified by the American Board of Pathology.

Clinical laboratory scientists are eligible to join ASCP as an affiliate or associate member. Presently, associate membership is limited to those who hold technical certification through ASCP. Students in AMA(CAHEA)-accredited programs also are eligible for affiliate membership.

The ASCP offers a variety of educational programs, including a number of workshops, seminars, and scientific meetings, to pathologists and clinical laboratory scientists. Attendance at meetings and educational materials produced by ASCP are offered to members at reduced fees. The monthly journal *Laboratory Medicine* is available to ASCP members.

American Association of Bioanalysts

Established in 1956, the American Association of Bioanalysts (AAB) represents independent community laboratories. Members include directors, managers, supervisors, and owners of community clinical laboratories.[4]

American Association of Blood Banks

Established in 1947, the American Association of Blood Banks (AABB) is a professional, nonprofit, scientific, and administrative association for individuals and institutions engaged in the many facets of blood banking and transfusion medicine. The mission of the AABB is to provide leadership in blood procurement, processing, and distribution and in all aspects of transfusion medicine and related fields, to promote the highest standards of care and service to patients, donors, other health care professionals, and the general public.

AABB member facilities are responsible for collecting nearly half of the nation's blood supply, and for transfusing more than 80% of it. More than 2,400 institutions (community and hospital blood banks and hospital transfusion services) and 8,000 individuals are members of the AABB, including physicians, scientists, medical technologists, administrators, blood donor recruiters, nurses, and public-spirited citizens working with blood banks. The AABB is also engaged in voluntary inspection and accreditation of blood banks and transfusion facilities, a service instituted in 1958.

In 1954, the first examination for blood bank technologists was administered by the American Society of Clinical Pathologist Board of Registry. It is significant that both the examination and the process for approving blood bank schools were the result of a cooperative effort between the Board of Registry and the AABB. Technologists working for 5 years in blood banking were eligible to take the examination. As of 1990, 13,786 individuals have been certified as specialists in blood bank technology.[3,5]

American Association for Clinical Chemistry

The American Association for Clinical Chemistry (AACC), an international, nonprofit society of clinical laboratory professionals, counts more than 9,000 members. The organization offers continuing education workshops and seminars, sponsors three scientific meetings each year, and publishes a newsletter, *AACC News*, and a journal, *Clinical Chemistry*.[4]

The American Association of Immunologists

The American Association of Immunologists (AAI) was founded in 1913. The purpose of the Association is to advance knowledge of immunology and related disciplines and to facilitate interchange of ideas and information among investigators in the various fields. AAI publishes a journal (*Journal of Immunology*), convenes an annual scientific meeting, and sponsors an advanced course in immunology.

The AAI promotes the interaction between clinicians and laboratory investigators. To assure that certifying examinations reflect the current state of knowledge in immunology, representatives of AAI interact with various groups engaged in certification activities.

Members receive a newsletter, reduced subscription rates to professional journals, and reduced registration fees to the annual meeting.

The Clinical Immunology Society (CIS) and Association of Medical Laboratory Immunologists (AMLI) are two other professional organizations whose members have an interest in immunology.

American Society of Allied Health Professions

The American Society of Allied Health Professions (ASAHP) was incorporated in 1967 as the Association of Schools of Allied Health Professions. The name was changed in 1973 to reflect the changes in membership and in the organization of the Society. The Association consists of individual members, colleges, universities, community colleges, vocational-technical centers, institutions, professional organizations representing specific health care disciplines, and a few interested corporate members. More than three million professionals and hundreds of institutions and organizations belong to ASAHP.[7]

The ASAHP mission is to contribute to the improvement of health care by enhancing the effectiveness of education for allied health professions. Its purposes are to provide leadership in education for schools, colleges, divisions, and departments of allied health professions and to serve as their representative and spokesman. It hopes to facilitate cooperation and communication among its diverse members, promote new allied health program development, and encourage research and study related to education of allied health professionals. Serving as both an information resource and as a catalyst for innovations in allied health education, the Society works continually to support both the professional community's demand for practice competencies and human resources and academia's demand for scholarship, teaching, and service. The society publishes a quarterly journal, *Journal of Allied Health*, and a newsletter, *Allied Health Trends*.

American Society of Hematology

The American Society of Hematology brings together hematologists throughout the Americas to expand the field of knowledge and improve the treatment of diseases of blood and blood-forming tissues.

The Society has more than 4,000 members in the United States, Canada, and Latin America. The vast majority of the members are board certified in Hematology.

The American Society of Hematology was founded in April, 1958, by 150 hematologists who met in Boston in response to the need for a forum for those involved in the broad array of sciences that deal with the study of blood. The Society has played an active role in bringing together clinical practitioners and scientists and in the development of hematology as a discipline through the establishment of standards for certification.

The Society sponsors an annual scientific meeting and symposia, and offers a number of educational opportunities. It publishes the journal *Blood* and a quarterly newsletter for Society members.

American Society for Microbiology

Founded in 1899, the American Society for Microbiology (ASM) is the oldest and largest single biologic science membership organization in the world, with more than 37,000 members in more than 22 disciplines throughout the United States and in most foreign countries. The American Academy of Microbiology sponsors three professional certification programs at the bachelor's, master's, and doctoral degree levels:

1. The National Registry of Microbiologists (NRM) offers programs at the bachelor's, master's, and doctoral levels. Certification is available as a Registered Microbiologist (RM(AAM)) in the areas of Clinical and Public Health Microbiology, Research and Development Microbiology, and Consumer Products and Quality Assurance Microbiology. Specialist certification (SM(AAM)) is available for microbiologists at the supervisory level in Public Health and Medical Laboratory Microbiology and in Consumer and Industrial Microbiology. The NRM is sponsored by the American Academy of Microbiology.

2. The American Board of Medical Microbiology (ABMM) offers certification at the doctoral level in Medical and Public Health Microbiology, Mycology, Virology, and Parasitology.

3. The American Board of Medical Laboratory Immunology (ABMLI) offers certification at the doctoral level in medical laboratory immunology.

Members receive a quarterly newsletter and a monthly news update, and are eligible for member rates for 10 journals published by ASM. Professional conferences and meetings are sponsored by ASM, and a job placement service is available to members. In addition, the Public and Scientific Affairs Board monitors legislative and regulatory issues.

Any person interested in microbiology who holds at least a bachelor's degree or equivalent experience in microbiology or related field is eligible to become a full member.

Clinical Laboratory Management Association

The Clinical Laboratory Management Association (CLMA) is a 5,200-member professional association dedicated to educating laboratory managers in business aspects of a clinical laboratory. CLMA members represent the management team (laboratory director, administrative director, and hospital administrator) of hospitals, independent laboratories, and industries that serve the clinical laboratory.[4]

College of American Pathologists

The College of American Pathologists (CAP) is the outgrowth of the separation of the educational, research, and socioeconomic activities from the scientific and business portions of ASCP. To be eligible for full membership, a physician must be a diplomate of the American Board of Pathology.[7]

International Association of Medical Laboratory Technologists

The International Association of Medical Laboratory Technologists (IAMLT) provides a mechanism for communication among the technologists of different countries. The association hosts an annual convention to which ASMT sends delegates. Membership is not open to individuals, but to organizations.[7, 10]

INFORMATION MATERIAL

Additional information is available from

American Association of Blood Banks (AABB)
American Association of Immunologists (AAI)
American Association for Clinical Chemistry (AACC)
American Medical Technologists (AMT)
American Society of Hematology (ASH)
American Society for Medical Technology (ASMT)
American Society for Microbiology (ASM)
American Society of Clinical Pathologists (ASCP)
Board of Registry of American Society of Clinical Pathologists
Committee on Allied Health Education and Accreditation (CAHEA)
International Society for Clinical Laboratory Technology (ISCLT)
National Accrediting Agency for Clinical Laboratory Sciences (NAACLS)
National Certification Agency for Medical Laboratory Personnel (NCA)

REFERENCES

1. American Society for Medical Technology: Bylaws and society regulations of the American Society for Medical Technology, revised June 1973.
2. Board of Registry: Procedures for examination and certification. Chicago, 1990, American Society of Clinical Pathologists.
3. Cherney PJ: Board of Registry newsletter. Chicago, American Society of Clinical Pathologists, Oct 1990.
4. FYI: Organization data listed. *Advance* 1990; 2:13.
5. Gupta GC: *Allied Health Directory 1990*, ed 18. Chicago, 1990, American Medical Association.
6. Karni K, Oliver JS: *Opportunities in Medical Technology Careers*. Lincolnwood, Ill, 1990, NTC Publishing Group.

7. Lindberg DS, Britt MS, Fisher FW: *Williams' Introduction to the Profession of Medical Technology*, ed 4. Philadelphia, 1984, Lea & Febiger.
8. Miller-Allan P: *Introduction to the Health Professions*. Monterey, Calif, 1984, Wadsworth.
9. Montgomery LG: A short history of the Registry of medical technologists of the American Society of Clinical Pathologists. *Am J Clin Pathol* 1970; 53:433.
10. Pohl S: The International Association of Medical Laboratory Technologists: A personal view. *J Med Tech* 1984; 1:578.

Chapter 8

Overview of Laboratory Departments

COAGULATION LABORATORY

Depending on the size of the clinical laboratory, hematology and coagulation may be considered the same department, located in the same room, or as separate, with individual or compartmentalized spaces, budgets, and personnel.

The coagulation department is concerned with laboratory testing associated with bleeding and clotting problems. Do you ever wonder what happens to stop the blood flow after you get a paper cut? The coagulation laboratory is interested in studying this process of *hemostasis*, or cessation of blood flow from an injured blood vessel. We know the clotting process involves a series of steps. The first step is the constriction (narrowing) of blood vessel walls. Platelets, small clotting particles, then move to the injury site and form a primary hemostatic plug. A series of reactions activate each of the 12 coagulation factors in a sequential fashion. This ultimately produces a stable fibrin clot. The fibrin clot eventually dissolves (*fibrinolysis*).

Hemostatic or coagulation disorders can be attributed to a variety of causes, such as low numbers of or poorly functioning platelets, a coagulation factor deficiency, a blood vessel disorder, or deficiency of factors associated with the fibrinolytic system. Perhaps the best-known bleeding disorder is hemophilia, caused by a deficiency of factor VIII. This sex-linked recessive genetic trait (which means it is located on the X [female] chromosome) most frequently affects males. Hemophiliacs show easy bruising, and have difficulty in stopping blood flow resulting from cuts, dental extractions, surgery, and the like. The coagulation laboratory assists the physician in diagnosing coagulation diseases and identifying the specific coagulation factor deficiency(s).

The two most commonly performed tests in the coagulation laboratory are prothrombin time (PT) and activated partial thromboplastin time (aPTT). These are screening tests, performed to identify potential "bleeders," and also are used to monitor anticoagulant therapy. After myocardial infarction (heart attack) or cerebrovascular accident (stroke), both due to blood clots, the patient may be given medicines that the public refers to as "blood thinners." These medications are actually anticoagulants, substances that prevent or slow the clotting process and help prevent future blood clots from forming. These patients must be monitored, because too large a dose of anticoagulant could lead to bleeding problems. The results of PT and aPTT are useful in monitoring these patients.

Almost all coagulation laboratories have at least one semiautomated instrument known as a fibrometer. This versatile instrument is the workhorse of the coagulation laboratory. Many clinical laboratories of any size also have more automated ("walk away") instruments for determining PT and aPTT. Once the patient specimen is prepared and placed in the sample tray, the technologist can walk away while the instrument performs the test. Most of these automated instruments enable the technologist to perform more sophisticated assays, such as individual factor(s) and fibrinogen assays.

Coagulation laboratories will continue to become more automated in the future. There will be an increasing need for automation that requires less personnel time and more accurate results.

The last few years have seen the development of coagulation tests that involve chromogenic substrates. The coagulation factors are, in fact, enzymes, that is, chemicals that help to make a chemical reaction. When these enzymes act on their beginning substance, known as a substrate, it turns the substrate into a different chemical. This resulting chemical is known as the end product. Thus we can indirectly measure the amount of enzyme (or coagulation factor, in this case) by measuring the amount of end product (usually indicated by a color change). In the future, the coagulation laboratory may be part of the chemistry department and use instruments similar to spectrophotometers.

Another exciting advancement in the past few years has been the use of fibrinolytic medications for patients who have had recent heart attacks or strokes. These medications help dissolve blood clots that have already formed.

HEMATOLOGY LABORATORY

The term "hematology" is derived from the Greek words *haima* (blood) and *logos* (study of or science of). Hematology is the study of blood. An adult has a total blood volume of approximately 6 L. Approximately 35% to 45% of the blood consists of formed elements, and 55% to 65% is plasma. Plasma consists primarily of water, with small amounts of salts, protein, and sugars. The formed elements are blood cells, and include erythrocytes (red blood cells), leukocytes (white blood cells), and platelets (actually, fragments of cells assisting in blood coagulation).

Blood has several functions. It carries nutrients and oxygen to tissues, carries away waste products, and serves as a natural defense against foreign materials such as bacteria and viral particles.

The primary screening test for blood abnormalities is the complete blood cell count (CBC), which determines the number of red blood cells (RBC), white blood cells (WBC), and platelets (plt) in a measured volume of blood, along with hemoglobin (Hgb) concentration and hematocrit (Hct; percent), and white blood cell differential (diff). The CBC is helpful in detecting anemias (in which there are too few red cells), leukemias (in which there are too many white cells), and infections.

A CBC in most clinical laboratories is performed using an automated instrument. Blood cell counters are available with a wide range of sophistication. These instruments may also do automated differential analyses, which identify the white cells as to type based on cell size, maturity, and nuclear and cytoplasmic staining characteristics. When done by hand, the technologist differentiates 100 white cells using a stained blood smear. Types of white cells include eosinophils, basophils, lymphocytes, monocytes, and segmented neutrophils. Cell count and differential analysis may be performed on other body fluids also, for example, cerebrospinal fluid (CSF), from the spinal cord; pleural fluid, from the lungs; peritoneal fluid, from the abdominal cavity; and synovial fluid, from the joints.

Two other laboratory tests frequently performed in the hematology laboratory are the reticulocyte count and erythrocyte sedimentation rate (ESR). Hemoglobin electrophoresis also may be done to confirm hereditary hemoglobin disorders, such as sickle cell anemia. Technologists may also work with bone marrow specimens. Bone marrow is located within the bone and contains precursors to circulating white blood cells, red blood cells, and platelets. A bone marrow is generally performed to help confirm the diagnosis of a particular anemia or leukemia.

Technologists who enjoy working in hematology generally like working with automated instrumentation, enjoy microscope work, and have good skills in cell identification.

The future of hematology lies in increasing and improving automated instrumentation. Fewer differentials will be done by hand. The use of new technology, such as flow cytometers, will expand in the future. Flow cytometers help differentiate and count cells on the basis of how a cell scatters a laser light beam. Flow cytometry will be used in diagnosing leukemias more specifically as to cell type and in obtaining more accurate WBC differentials. This will significantly help the physician determine the right treatment regimen for a particular blood disorder.

URINALYSIS LABORATORY

The urinalysis laboratory generally is a section of the hematology laboratory, or may be a separate department in larger laboratories. Laboratory professionals are responsible for routine testing on random urine specimens. Urinalysis is the oldest laboratory test still performed.

Routine urinalysis is a valuable tool for the detection of disease related to the kidney, such as kidney or bladder infections. It is useful in diagnosing and monitoring metabolic diseases such as diabetes, phenylketonuria (PKU), and cancers of the urinary tract.

A routine urinalysis consists of three parts:

1. The physical characteristics of the specimen are described including an evaluation of the clarity (clear or cloudy) and color of the sample. The specific gravity or concentration of the specimen is measured and reported. These observations of the urine can be important in the overall diagnosis of the patient.

2. The urine specimen is chemically tested. The process is performed using a plastic strip containing nine or 10 pads impregnated with chemicals that react with medically important substances in the urine, such as glucose, ketones, blood, protein, bilirubin, urobilinogen, nitrites, and leukocyte esterase. pH and specific gravity also are measured by this technique.

3. The urinary sediment is examined microscopically. The technologist looks for cells, casts, or crystals that might be due to a disease process in the renal system. Important cells are

red and white blood cells; they might indicate bleeding or infection. Casts are made up of protein and are formed in the renal tubules; they usually indicate a serious kidney problem. Crystals in the urine may be the result of certain metabolic disorders; they can lead to the development of kidney stones.

BLOOD BANK (IMMUNOHEMATOLOGY LABORATORY)

Have you ever wondered what happens to a unit of blood that you or a family member has donated? The blood will be used in a transfusion, in which a patient will receive the unit of blood. The American Red Cross is the largest organization having responsibilities for obtaining blood from donors and supplying this blood to hospitals.

A donor center is responsible for ensuring that the blood supply is safe. This is done in a number of ways. A medical history on all donors is obtained prior to donation. The donor's blood pressure, temperature, pulse rate, and hemoglobin are measured; this is done to help assure that the donor will not suffer any ill effects from the donation. Once collected, the blood is typed. Blood may be typed as O, A, B, or AB. Blood also is typed according to Rh (D factor), and is described as positive (having the Rh_oD factor) or negative (lacking the Rh_oD factor). In addition, the blood is screened for unusual antibodies and for hepatitis viruses and for human immunodeficiency virus (HIV), the etiologic agent of acquired immune deficiency syndrome (AIDS). (NOTE: You *cannot* get AIDS by donating blood.)

Once the unit of blood arrives at the hospital, the ABO and Rh typings are confirmed, and the blood is refrigerated until needed. When an order for a unit of blood arrives in the hospital blood bank, a sample of blood must be drawn from the patient to determine ABO and Rh type and to screen for unusual antibodies. As in all clinical testing, patient sample identification is essential throughout the process. Once the patient ABO and Rh type is determined, a "compatible" unit of blood is found. Generally, a unit of blood of the same ABO and Rh type is chosen. Compatibility testing is performed, mixing a small amount of patient serum (the liquid portion of the blood) and a small amount of red blood cells from the donor unit in a test tube. If agglutination ("clumping") occurs in the test tube, it is likely this will happen in the patient's blood vessels once the unit is transfused. The unit of blood is considered "incompatible," and must not be given to the patient tested. The technologist must find a unit of blood that does not cause clumping.

Once compatibility testing is completed, the unit of blood may be released for transfusion. Again, maintaining correct patient identification and donor bag identification is essential. A mistake can result in patient death.

Most of the testing done in the blood bank is based on antigen-antibody reactions. Antigens are specific proteins, in this case attached to the red and white blood cells. They help make each person's blood cells unique. For example, some people have A antigen (making their blood type A), some have B antigen (type B), some have both antigens (type AB), and some lack both A and B antigens (type O). In most blood group systems, we generally do not have antibodies to antigens that we have on our red cells. However, if a person is exposed through transfusion or pregnancy to blood of a different type, he will recognize this transfused material as "foreign" and respond by making antibodies, proteins that will react with the antigen. When a unit of blood is incompatible, it generally means that there is an antigen on the donor red cells to which the patient has responded by making an antibody. If this incompatible unit of blood is transfused, it can cause a transfusion reaction and possible death.

The blood bank also is involved in component therapy. Many patients do not require a whole unit of blood, but one specific part of it. For example, platelet components can be prepared for patients with low platelet counts, and cryoprecipitate can be prepared for a patient with hemophilia. Numerous patients can be helped from one unit of donor blood.

Blood bank technologists also identify antibodies. Each patient specimen and each unit of donor blood is screened for the presence of antibodies. If a screen indicates the presence of an antibody, the technologist then performs a more specific procedure using a panel of red blood cells with known antigen specificity. Specific antibody(ies) are then identified.

Blood bank technologists also test prenatal, maternal, and postnatal specimens. Just as a foreign antigen from a transfused unit of blood can stimulate antibody production, a mother can make an antibody to an antigen that she lacks but her fetus has. Normally the blood systems of the mother and fetus remain separate. However, during the birthing process some of the fetal cells may enter the mother's blood, stimulating it to make antibodies. During a subsequent pregnancy or transfusion, when the mother is again exposed to the antigen, her antibodies are ready to "attack" the foreign cells. Thus, even before birth, the fetus' cells are being destroyed. Without therapy the fetus could be born very ill, or dead. Specimens from the mother, collected prior to giving birth, are used to determine if her blood contains antibodies, and if so, she and the fetus are monitored closely throughout the pregnancy.

The blood bank is not highly automated. At large donor centers, ABO and Rh typings may be done by automated instruments, because of large work volume, but most clinical laboratories rely on highly skilled technologists using test tubes and pipets. Record keeping in blood banks is becoming more automated. Computers keep track of quality control, blood inventory, and test results.

Blood bank technologists must be meticulous. They must be accurate record keepers and able to function in highly stressful situations. They usually enjoy the challenge and problem-solving activities that a position in blood banking presents.

CHEMISTRY LABORATORY

The clinical chemistry laboratory is responsible for testing body fluids for several hundred different chemical substances. These fluids most commonly are blood serum, urine, or (CSF). The 50 or so common tests performed in the chemistry laboratory include determination of serum glucose and total protein concentrations.

The serum glucose test is used to diagnose and monitor diabetes. The analysis can be performed with an enzyme, glucose oxidase, that reacts specifically with glucose and no other carbohydrate. The enzyme converts glucose to glucuronic acid and hydrogen peroxide. The enzyme peroxidase and a chromogenic oxygen acceptor react in the second step. The peroxidase, oxygen acceptor, and the peroxide create a colored product. This colored compound is measured using a spectrophotometer. The degree of color change is proportional to the concentration of glucose in the sample.

Many different types of chemicals aid in the diagnosis of disease. Creatinine is a waste product excreted by the kidneys. Analysis of this substance can be used to diagnose renal disease. The study of enzymes in cells of different body tissues also can help identify damage or disease of specific organs. Lactate dehydrogenase (LDH) and creatine kinase (CK) are enzymes used to identify heart disease and to monitor patients after a heart attack.

Most routine chemistry testing is performed on extremely sophisticated robotic, computerized, electronic instruments. This type of automation allows the technologist to efficiently and accurately analyze a large number of specimens. Most of these instruments are found in the general or routine chemistry section.

Other divisions within the chemistry laboratory use instruments that are not so automated but are as sophisticated. The toxicology laboratory technologist helps monitor therapeutic drug levels and identifies drugs of abuse. The gas chromatograph–mass spectrometer allows the technologist to identify and quantitate many types of drugs.

The individual that would enjoy clinical chemistry laboratory work will like working with electronic, computerized instruments. Some areas challenge the technologist more frequently in problem-solving situations, such as drug identification. The technologist must be able to evaluate the results produced by the instruments. To do this requires knowledge of the pathology of disease states and how the disease process affects the chemistry tests they have performed.

Clinical chemistry laboratories vary in complexity depending on the size of the hospital and the type of institution (e.g., private reference laboratory, physician office laboratory). A large chemistry laboratory may be subdivided into smaller sections, for example, a general chemistry section for most routine tests, a toxicology and therapeutic drug monitoring section, and ligand laboratory section for radioimmunoassay (RIA) and enzyme immunoassay (EIA).

A number of professional organizations function to promote research and education in the field of clinical chemistry. These include the American Association for Clinical Chemistry, American Board of Clinical Chemistry, American Board of Forensic Toxicology, American Chemical Society, American Institute of Chemists, American Society for Biochemistry and Molecular Biology, Association of Clinical Scientists, and the National Academy of Clinical Biochemistry.

Immunology/Serology

Immunology is the study of the immune system. The function of the immune system is to protect the body from damage caused by invading microorganisms, such as bacteria, viruses, fungi, and parasites. This defensive function is performed by white blood cells (lymphocytes, monocytes, and other leukocytes) distributed throughout the organs of the body. The various cells interact with each other, producing a coordinated immune response directed at eliminating the pathogen (substance causing a disease) or minimizing the damage it causes.[1]

The immunology/serology component of the clinical laboratory deals with three major areas: immunopathology, diagnosis of infectious disease, and quantitation of serum components utilizing immunologic methods.

Just as in the blood bank, many of the immunologic laboratory tests are based on antigen-antibody reactions. Protein substances, known as antigens, are found as components of bacteria, viruses, and even toxins on the cell surface of blood cells and tissue cells. When foreign material is introduced, such as a virus or transplanted tissue, the body reacts by making antibodies to eliminate it.

Patients with AIDS show an abnormal response when foreign matter is introduced. The immune system of AIDS patients often is not capable of destroying this foreign material. These patients are then more susceptible to a variety of infections (e.g., pneumonia) and cancers (e.g., Kaposi sarcoma).

The diseases of the immune system are numerous. Most immunodeficiencies are either congenital (hereditary) or, as in the case of AIDS, acquired. Another immune system disorder, multiple myeloma, is a malignant disease associated with overproduction of immunoglobulins.

Allergies are responses induced through the immune system. Many individuals suffer from allergies, which can be graded from annoying to life threatening.

Other immunologic diseases result when the body makes antibodies against its own cells. Examples are rheumatoid arthritis and lupus erythematosus.

The diagnosis of these immune system disorders relies on the ability of the technologist to differentiate normal and abnormal levels of immune cells and serum components. In certain cases, the functions of immune cells such as lymphocytes, monocytes, and leukocytes are determined. The laboratory makes use of routine techniques such as precipitation or agglutination reactions to quantitate and identify serum components. Simple tests exist for detecting diseases such as infectious mononucleosis, pregnancy, rheumatoid arthritis, and syphilis. In addition, sophisticated instrumentation such as nephelometry, radioimmunoassays, and enzyme immunoassays may be used for quantitation of immunoglobulins or other factors. Fluorescent microscopy, monoclonal antibody tests, and flow cytometry also are used in some immunology tests.

The diagnosis of infectious diseases often is complemented by the serology laboratory. Rapid diagnostic procedures for septic sore throat ("Strep" throat) and other common infectious diseases have been developed that rely on agglutination reactions, using specific antisera containing enzyme-labeled antibodies. The microbiology laboratory also may request sophisticated immunologic tests for the identification of unusual microorganisms.

The quantitation of serum proteins such as immunoglobulins (antibodies), albumin, and acute phase proteins are routine procedures in the immunology laboratory. Simple techniques such as radial immunodiffusion (RID) often are used when the volume of tests does not warrant the acquisition of expensive instrumentation.

In hospitals that do not have immunology laboratories, the functions of the immunology laboratory may be done in the chemistry, immunochemistry, or microbiology laboratory or in some combination of departments. Quantitation of immune cells can be done in the hematology, tissue typing, or blood banking departments.

The distinguishing feature of the immunology laboratory is the limited amount of "stat" or emergency procedures. Laboratory work can be better organized and structured. Thus some of the stress associated with the clinical laboratory is lessened in this department. Workers in this department must remain current in the field. Immunology testing has vastly expanded during the 1980s, and promises to do so in the 1990s. Laboratory personnel also may be expected to develop, apply, and evaluate new tests and methods for the immunology department. An interest in research and development is helpful to technologists in immunology.

The immunology laboratory of the future in some aspects is here today. Since the development of monoclonal antibody techniques, the number of assays and techniques has increased enormously. Sophisticated instrumentation will continue to be developed to utilize the ever expanding list of antibodies to detect smaller and smaller concentrations of serum constituents.

Many immunologic hormones, lymphokines, and monokines have been isolated, identified, and manufactured in the last few years. These substances currently are being used to treat many immunologic diseases. It is likely that the immunology laboratory of the future will offer tests to monitor the amounts of these substances as well as their effectiveness.

MICROBIOLOGY

The specialty of clinical microbiology deals with the identification of microorganisms that can cause disease. Microbiology is one of the older departments in the clinical laboratory. We have known for a long time that infectious substances can transmit disease. This was graphically evident with spread of bubonic plague in Europe during the Middle Ages. Pioneers in the field of microbiology (e.g., Pasteur and Koch) were clinical microbiologists interested in identifying important pathogens (substances causing disease) in humans. The use of the microscope was critical to these early investigators.

The work in clinical microbiology is varied. Depending on the size of the hospital, the microbiology laboratory may be involved in the isolation and identification of bacteria (bacteriology), viruses (virology), fungi (mycology), and parasites (parasitology). Most of the work focuses on bacteriology.

In health, the human body is host to bacteria that do not cause disease. These bacteria are referred to as "normal flora." Distinguishing pathogenic microorganisms from the normal flora makes the job of a clinical microbiologist similar to that of a detective investigating a crime, looking for a suspect.

The routine microbiology laboratory relies on selective and differential media to enhance the growth of suspected pathogens. Media contain a variety of substances, such as nutrients, salts, or blood, that facilitate bacterial growth. Once a suspected pathogen is found, its identification is confirmed by biochemical or immunologic tests. Identification of most routine pathogens can be confirmed in 1 or 2 days. However, organisms such as *Mycobacterium tuberculosis*, which causes tuberculosis, require weeks to identify because of their extremely slow growth rate. Some microorganisms may require complex media just to grow. The identification of legionnaires' disease was not accomplished until a media was developed to grow the organism *(Legionella pneumophila)*. Other organisms, such as *Treponema pallidum*, the causative agent of syphilis, still cannot be cultured on artificial media, although the disease has existed for centuries. Some pathogens, such as *Clostridium perfringens*, the causative agent of gas gangrene, are killed by air and must be cultured under anaerobic (lacking oxygen) conditions. The clinical microbiologist must be aware of these growth requirements when trying to identify a pathogen.

Identification of pathogenic viruses, such as herpesvirus and hepatitis virus, requires the use of tissue culture cells and immunologic methods. Identification of pinworms can be routine, but the identification of other parasites, such as those responsible for malaria, is more demanding. Mycology involves the blending of macroscopic observations (e.g., growth rate, colony structure, color), microscopic observations, and biochemical methods to identify pathogenic fungi. Like the tuberculosis microorganism, fungi grow slowly; identification and isolation can take weeks or months.

In addition to identification of the pathogen, the clinical microbiologist is also called on to identify antibiotics that will be most effective in killing a microorganism. Once a pathogen is identified, the microorganism is tested against a battery of antibiotics to determine its susceptibility and the dose of the drug needed to be effective. Many microorganisms are becoming resistant to certain antibiotics. Most strains of *Neisseria gonorrhoeae* have become resistant to penicillin in recent years. Antibiotic susceptibility information and the identification of pathogens aids the physician in diagnosis and treatment.

In recent years, rapid screening tests have been developed to identify routine pathogenic microorganisms. Immunologic tests that use specific monoclonal antibodies are now used in the identification of "Strep throat" and gonorrhea. These rapid screening tests take 10 to 15 minutes and can be performed in the physician's office while the patient waits for the results. Other new tests rely on DNA probes. Every organism, including humans and bacteria, has a unique DNA sequence. Scientists have been able to identify DNA sequences for a number of microorganisms, and have developed tests to detect specific DNA sequences found in pathogenic bacteria. Because of the sensitivity of the tests, pathogens can be identified directly from clinical specimens without culturing. However, the costs of these DNA probe methods at present are prohibitive. DNA probes are now used for microorganisms that are difficult to culture or require long incubation times. As the techniques become more cost effective, DNA probes will be used more commonly.

The distinguishing feature of the microbiology laboratory is the reliance on the technologist's knowledge, skill, and intuitive reasoning to identify suspect pathogens. Although the use of instruments and automation is increasing in the microbiology laboratory, they are used to complement the technologists, not replace them.

Many professional organizations and societies certify individuals in microbiology. The American Society of Clinical Pathologists and the National Certification Agency for Clinical Laboratory Sciences certify individuals with training and work experience at the technician, technologist, and specialist rank. The American Society for Microbiology certifies individuals from the technologist to the director level. Requirements can vary from a B.S. degree for a technologist to a Ph.D. or M.D. degree for a director. Opportunities are available in clinical research and pharmaceutical industries at all levels. Although a number of pathogens were identified years ago, the isolation and identification of new pathogens, such as the AIDS virus, makes the work of the microbiologist stimulating and challenging.

Acknowledgment

Special thanks to Clifford Renk, Ph.D., for writing the sections on Immunology and Microbiology, and to Gary Hammerberg, Ed.D., MT(ASCP), for writing the sections on Urinalysis and Chemistry.

REFERENCE

1. Male D: *Immunology: An Illustrated Outline.* New York, Gower Medical Publishing, 1986.

Chapter 9

Educational and Clinical Requirements for Clinical Laboratory Technicians and Clinical Laboratory Scientists

The educational criteria for a clinical laboratory science program state that the curriculum must:

1. Meet all college or university degree (A.D. or B.S.) requirements. This usually includes general education courses.
2. Meet all requirements for a major and minor in the sciences.
3. Include all preprofessional and professional requirements for national certification examinations and state licensure examinations.

CLINICAL LABORATORY TECHNICIAN, MEDICAL LABORATORY TECHNICIAN

Education

Educational requirements to become a clinical laboratory technician (CLT) or medical laboratory technician (MLT) include a high school diploma, with a solid foundation in high school sciences (biology, chemistry, mathematics, and computer science), plus one of the following:

1. A program leading to an associate degree in medical or clinical laboratory technology (MLT-AD or CLT-AD) or certificate (MLT-C).
2. An associate degree in a basic science, such as biology and chemistry, followed by extensive clinical experience in an approved laboratory.
3. A certificate of completion from a military program. This is a 50-week course in medical laboratory techniques including laboratory experience.[3, 10]

Two educational curricula are available to complete a CLT or MLT program. The MLT certificate (MLT-C) program, usually 12 to 15 months, is offered through vocational schools, technical schools, and hospitals. The curriculum includes medical ethics and conduct, basic laboratory solutions and media, medical terminology, basic elements of quality control, blood collecting techniques, basic microbiology, serology, hematology, and immunohematology. A clinical laboratory practicum in a hospital or clinical laboratory is included in the MLT-C program.[7]

The associate degree (MLT-AD or CLT-AD) programs are conducted in junior or community colleges, 2-year divisions of universities and colleges, and in other institutions that grant associate degrees. Courses are taught on campus and in affiliated hospital(s). The clinical courses on campus include application of basic principles as commonly used in the diagnostic laboratory. The teaching laboratory on campus focuses on general knowledge and basic skills, understanding principles, and mastering procedures of laboratory testing. Technical instruction, on campus and in clinical practicums, includes procedures in hematology, serology, chemistry, microbiology, and immunohematology. The associate degree program usually encompasses 2 academic years.[6, 7]

It is suggested that the MLT-AD route be taken if one plans eventually to pursue a B.S. degree and certification as a clinical laboratory scientist. It sometimes is difficult to transfer credits earned from a vocational school to a college or university.[7]

More than 200 community colleges and approximately 40 hospitals offer CLT or MLT programs. A combination of formal education plus clinical training is required in a program accredited by the Committee on Allied Health Education and Accreditation (CAHEA) in cooperation with the National Accrediting Agency for Clinical Laboratory Sciences (NAACLS) and the Accrediting Bureau of Health Education Schools (ABHES).[1, 3, 9, 13]

Persons interested in a clinical laboratory career should carefully select an educational program. Prospective employers (hospitals and independent laboratories) may have preferences as to program accreditation. Educational programs should provide information about the kinds of jobs obtained by graduates, educational costs, how long the educational program has been in operation, instructional facilities, and faculty qualifications.[13]

Certification

To ensure that laboratory workers are competent, the Board of Registry of the American Society of Clinical Pathologists (ASCP) and the National Certification Agency for Medical Laboratory Personnel (NCA) offer national certification examinations to students who meet academic and clinical education requirements. Those who pass the examination for medical laboratory technicians may use the initials MLT(ASCP) after their names; those who pass the NCA exam may use the initials CLT. Eligibility for certification examinations is determined by clinical and educational experience.[3]

Requirements for Clinical Laboratory Technician Generalist Examination

For admission to the CLT generalist examination (NCA), applicants must satisfy one of four requirements[11]:

1. Graduation from a CLT program (or equivalent) accredited by an agency recognized by the U.S. Office of Education or approved by a state government agency.
2. Possession of a certificate of Military Laboratory Specialist MOS92B30, NEC 8417, NEC 8506, AFSC 92450, or AFSC 92470.
3. Possession of credentials from a foreign country attesting to clinical laboratory training and experience consistent with or superior to that for requirements 1, 2, or 4.
4. Four years of full-time work experience encompassing the four major disciplines of laboratory practice (clinical chemistry, hematology, microbiology, and immunohematology) within the last 7 years.

Requirements for Medical Laboratory Technician Examination

For admission to the MLT (ASCP) examination, applicants must satisfy one of six requirements[5]:

1. Associate degree or at least 60 semester hours (90 quarter hours) of academic credit from a regionally accredited college or university, including successful completion of a CAHEA-accredited MLT program and courses in biology and chemistry.
2. Associate degree or 60 semester hours (90 quarter hours) of academic credit from a regionally accredited college or university, including 6 semester hours (9 quarter hours) of chemistry, 6 semester hours (9 quarter hours) of biology, and Clinical Laboratory Assistant (CLA(ASCP)) certification. (NOTE: The CLA examination is no longer available.)
3. Thirty semester hours (45 quarter hours) of academic credit from a regionally accredited college or university, including course work as described in 2 above, in addition to successful completion of a CAHEA-accredited MLT-C program or successful completion of an advanced military medical laboratory specialist program.
4. High school graduation (or equivalent) plus successful completion of a CAHEA-accredited MLT-C program and, in addition, 1 year of full-time acceptable clinical laboratory experience within the last 10 years.
5. High school graduation (or equivalent) plus successful completion of an advanced military medical laboratory specialist program and, in addition, 1 year of full-time acceptable clinical laboratory experience within the last 10 years.
6. Associate degree or at least 60 semester hours (90 quarter hours) of academic credit from a regionally accredited college or university, including coursework as described in 2 above, and 5 years of full-time acceptable clinical laboratory experience in blood banking, chemistry, hematology, microbiology, immunology, and clinical microscopy within the last 10 years.

CLINICAL LABORATORY SCIENTIST, MEDICAL TECHNOLOGIST

Education

Educational requirements to become a clinical laboratory scientist (CLS) or medical laboratory technologist (MT) include a high school diploma, with courses in chemistry and biology, plus one of the following:

1. Completion of a 4-year CAHEA-accredited program leading to a bachelor's degree in clinical laboratory science or medical technology (3 + 1 or 2 + 2 program).
2. Completion of a bachelor's degree, including courses that meet the basic science requirements. Some MT and CLS programs require a college degree for admission (4 + 1 program).
3. Completion of a bachelor's degree in a biologic, physical, or chemical science plus appropriate laboratory experience that qualifies one for employment as a chemistry, hematology, or microbiology technologist.[10]

A combination of formal education (B.S. degree) plus clinical education in an MT or CLS program accredited by CAHEA is required. These 420 accredited programs are offered by 83 colleges and universities, 59 medical centers, and 278 hospitals. Hospital programs generally are affiliated with colleges or universities and lead to a bachelor's degree, although a few hospital programs require a bachelor's degree for entry.[1, 4, 6]

Academic Coursework

Bachelor's degree programs in clinical laboratory science include substantial course work in chemistry, biologic sciences, microbiology, and mathematics. The final component of course work is devoted to acquiring the knowledge and skills used in the clinical laboratory. In addition to basic science, many programs offer or require course work in management, business, and computer applications.

The curriculum varies from institution to institution. However, some courses are common to all curricula, such as biology, inorganic chemistry, organic chemistry, and mathematics. Elements of a liberal arts education are usually included, such as English, history, speech, and literature.

Types of Programs

Most CLS programs are one of three types: 3 + 1, 2 + 2, or 4 + 1. The first number refers to the years spent in college; the second number refers to the years of clinical training. Thus a 3 + 1 program requires 3 years of college, generally 90 semester hours, prior to 1 year of clinical experience. The senior year is spent at an affiliated hospital. In a 4 + 1 program the student completes 4 years of college, generally already having a B.S. degree prior to the 1 year of clinical training. Other formats are 2 + 2, 2 + 1, and 3½ + ½. The sample CLS curriculum as shown in Table 9–1 is a 3½ + ½ program. Students complete 7 semesters of academic course work and spend the last 6 months in clinical rotation.[8]

All CLS programs have a structured laboratory component that includes instruction relating theory to practice in hematology, clinical chemistry, microbiology, immunology, and im-

TABLE 9–1.
Sample 2½ + ½ CLS Curriculum: Eastern Michigan University Clinical Laboratory Sciences Program

Clinical track
Freshman
  Fall   Suggest: 15–18 cr
    MTH 105   College Algebra *or* waiver   3 cr
    BIO 105   Biology & Human Species *or* waiver   4 cr
    CHM 121, 122   General Chemistry I lecture and lab   5 cr
    CLS 101   Intro to Clinical Laboratory Sciences   1 cr
    AREA I Elective   One course in Eng Comp *or* Foreign Language   3–5 cr
  Winter   Suggest: 16–17 cr
    CHM 123, 124   General Chemistry II lecture and lab   4 cr
    AREA I   One course Fund of Speech   2–3 cr
                Physical Education   1 cr
    CSC 136, 137   Computer Science   3 cr
    AREA II   Science or Technology (not BIO or CHM)   3 cr
    AREA III   Anthropology, Economics   3 cr
Sophomore
  Fall   16 cr
    CHM 270   Organic Chemistry lecture   4 cr
    CHM 271   Organic Chemistry lab   1 cr
    BIO 301   Genetics   3 cr
    ZOO 201   Human Anatomy and Physiology I   3 cr
    AREA IV   Literature or Foreign Language   3 cr
    CLS 200   Clinical Laboratory Techniques   2 cr
  Winter   16 cr
    CHM 281   Quantitative Analysis   4 cr
    MIC 329   General Microbiology   5 cr
    PLS 112, 113, 202   Government   3 cr
    ZOO 202   Human Anatomy and Physiology II   3 cr
                Physical Education   1 cr
Junior
  Fall   17 cr
    CLS 335   Clinical Immunology   4 cr
    CHM 351   Intro to Biochemistry   3 cr
    CHM 352   Biochemistry lab   2 cr
    AREA IV   Fine Arts   3 cr
              Literature or Foreign Language   3 cr
    CLS 310   Urinalysis/Body Fluids   2 cr
  Winter   16 cr
    CLS 432   Clinical Microbiology   4 cr
    CLS 402   Clinical Chemistry   3 cr
    CLS 307   Intro to Hematology   3 cr
    CLS 270   Elementary Statistics   3 cr
    NUR 320   Pathology   3 cr
Senior
  Spring/summer   9 cr
    ZOO 462   Parasitology   3 cr
    AREA III   History   3 cr
    AREA I   Group I Elective   3 cr
  Fall   16 cr

(Continued.)

TABLE 9-1 (cont.).

CLS 401	Management/Education/QA	4 cr
CLS 434	Adv Immunohematology	3 cr
CLS 407	Adv Hematology	3 cr
AREA III	History	3 cr
AREA IV	Religion or Philosophy	3 cr
Winter		16 cr
Clinical Rotation, 15 wk		15 cr
CLS 457	Clin Methods Practicum	1 cr
Spring		6 cr
Clinical Rotation, 1-8 wk		
CLS 459	Clin Mycology Practicum	1 cr
CLS 450, 452, 454, 456 (one 5-wk rotation)		5 cr
Total		138-142 cr

munohematology. The program must culminate in a baccalaureate degree for those students not already possessing the degree.[6]

The 3 + 1 and 4 + 1 Programs.—Many programs are arranged to include 3 years of undergraduate work followed by 1 year of intensive clinical education at an affiliated school of medical technology. The school usually is located within a hospital or university medical center. The clinical practicum must be a cooperative effort with the university, and according to NAACLS *Essentials,* there must be a signed affiliation agreement between the two institutions.[8] On completion of the 1-year clinical education component, a bachelor's degree is awarded, and students are eligible to take national certification examinations.

In California, the majority of clinical laboratory programs follow the 4 + 1 model: 4 years in college followed by a 1-year practicum or "internship." A baccalaureate degree is obtained before beginning the clinical practicum. The degree may be in any area as long as the student meets the pretechnical requirements of chemistry, biology, and mathematics. Some universities may describe the medical technology curriculum in their catalog, but do not offer specific preprofessional courses.[7, 8]

In both the 3 + 1 and 4 + 1 programs the student is responsible for obtaining clinical laboratory experience. Students should verify that appropriate affiliation agreements exist with hospitals and determine past student success in obtaining internship experience. Some universities and colleges facilitate these matters better than others.

Hospital-based CLS programs take anywhere from 9 to 12 months, and should be accredited through CAHEA. All accredited CLS and CLT programs are listed in the *Allied Health Education Directory.*[6]

Clinical rotation schedules vary from program to program. Generally, one or two students are placed in a particular laboratory department at a time. On completion of training in that department the student rotates into another department. One disadvantage to this system is that the lecture series may not correlate with the student laboratory rotation. A student may be attending blood bank lectures while training in the chemistry department, for example.

On completing the clinical practicum, the student is awarded a certificate or diploma in clinical laboratory science or medical technology.

The 2 + 2 Integrated Program.—In colleges and universities, which themselves are accredited in clinical laboratory science or medical technology, students usually spend 2 years in a preprofessional program and 2 years in a professional program. The first 2 years are spent in completing prerequisites in general chemistry, general biology, organic chemistry, mathematics, and microbiology, in a curriculum resembling that of premedicine. During the junior year students complete course work in computer science, statistics, immunology, anatomy and physiology, microbiology, and begin taking preclinical (introductory) laboratory courses in clinical chemistry, hematology, microbiology, and immunohematology. Student laboratories are generally conducted on campus. The students then complete basic rotations at the affiliated hospital(s). Because a campus student laboratory generally cannot include all types of equipment and testing available in a diagnostic laboratory, the hospital site enhances the campus laboratory experience. Students may then return to campus during the senior year and take management, education, and advanced clinical courses. This is known as the 2 + 2 integrated program.[7, 8]

Academic faculty usually are responsible for supervising the clinical experience, but the actual technical instruction is done by laboratory personnel employed at the clinical site. Students are evaluated by performance on written tests and laboratory practicals (cognitive knowledge), technical performance (psychomotor skills), and attitudinal (affective) behaviors.

Several advantages exist when preprofessional and professional course work is integrated with clinical rotations: (1) Students are able to apply what they have learned in the classroom to the diagnostic laboratory shortly after they have completed the introductory course. (2) Students know before their junior year that they have a placement in a clinical affiliate. Once accepted into the junior year, the student can be assured that there will be an opening at an affiliated hospital for training. (3) The campus laboratory experience is coordinated with the lecture. Students do not work in a laboratory department before receiving theory information. (4) Instruction is given by academic faculty who usually have training in educational methods and clinical laboratory science.

There are two disadvantages of the 2 + 2 curriculum: (1) Since some of the course work has been taken during the junior year, information may be forgotten before taking the certification examination, possibly 2 years later. (2) It is difficult to develop a schedule for students who wish to attend school only part-time.[7]

Studies indicate that there is no difference in performance on certification examinations on the basis of program type. Some employers state that after 6 months there is no difference in work performance, although initially the 2 + 2 student may be a little slower and less organized.[8]

CAHEA PROGRAM

CAHEA-accredited programs in clinical laboratory science are required to accept only those students who have completed the following academic prerequisites[6]:

1. *Biologic science:* A minimum of 16 semester hours (24 quarter hours), including immunology and a full course in microbiology, which should include a laboratory session. The course work in immunology may be included as a separate course or as part

of another course. These courses must be applicable toward a major in biology or medical technology. Survey courses are not accepted.
2. *Chemistry:* A minimum of 16 semester hours (24 quarter hours), including organic or biochemistry, either of which should include a laboratory session. These courses must be applicable toward a major in chemistry or medical technology. Survey courses are not accepted.
3. *Mathematics:* A minimum of one course in college mathematics, such as statistics, calculus, or algebra. Courses in remedial mathematics are not accepted.

NOTE: The organic or biochemistry course and the microbiology course must have been taken within the past 7 years or the course work must be updated before entering a medical technology program. Most accredited CAHEA programs have incorporated these requirements into the program curriculum.

Certification

To ensure that laboratory workers are competent to perform high-quality laboratory tests, the Board of Registry of ASCP and the NCA offer national certification examinations, which students may take after meeting academic and laboratory education requirements. Those who pass the ASCP examination may use the initials MT(ASCP) after their names; those passing the NCA examination use CLS.[4]

Requirements for Clinical Laboratory Scientist Generalist Examination

For admission to the CLS generalist examination (NCA), applicants must satisfy one of five requirements[11]:

1. Graduation from a structured education program in clinical laboratory science (medical technology or equivalent) that culminates in a baccalaureate degree and includes clinical experience in each of the four major disciplines of laboratory practice (clinical chemistry, hematology, microbiology, and immunohematology).
2. Completion of a baccalaureate program that includes a minimum of 36 semester hours or the equivalent quarter hours in the biologic and physical sciences plus completion of a clinical laboratory program (medical technology or equivalent) accredited by an agency recognized by the U.S. Office of Education or approved by a state governmental agency.
3. Completion of a baccalaureate program that includes 36 semester hours or the equivalent quarter hours in the biologic and physical sciences plus 2 years of full-time laboratory work experience including a minimum of 4 months in each of the four major disciplines of laboratory practice within the last 5 years.
4. Completion of 60 semester hours or the equivalent quarter hours of college course work including 36 semester hours or the equivalent quarter hours in biologic or physical sciences plus 4 years of full-time work experience within the last 10 years including a minimum of 6 months in each of the four major disciplines of laboratory practice.

5. Certification as a Registered Technologist (RT) by the Canadian Society of Laboratory Technologists plus 6 months of full-time laboratory experience in each of the four major disciplines of laboratory practice or certification as an Advanced Registered Technologist (ART) by the Canadian Society of Laboratory Technologists.

Requirements for Medical Technologist Examination

For admission to the MT(ASCP) examination, applicants must satisfy one of four requirements[5]:

1. Baccalaureate degree from a regionally accredited college or university, including courses in biologic science, chemistry, and mathematics, and successful completion of a CAHEA-accredited Medical Technology program.
2. MLT(ASCP) certification and a baccalaureate degree from a regionally accredited college or university, including 16 semester hours (24 quarter hours) of biologic science (with 1 semester in microbiology), 16 semester hours (24 quarter hours) of chemistry (with 1 semester in organic or biochemistry), and 1 semester (1 quarter) of mathematics, and 3 years of full-time acceptable clinical laboratory experience in blood banking, chemistry, hematology, microbiology, immunology, and clinical microscopy within the last 10 years.
3. CLA(ASCP) certification and a baccalaureate degree from a regionally accredited college or university, including course requirements as listed in 2 above, and 4 years of full-time acceptable clinical laboratory experience as described in 2 above.
4. Baccalaureate degree from a regionally accredited college or university, including course work as described in 2 above, and 5 years of full-time acceptable clinical laboratory experience as described in 2 above.

APPLYING TO A LABORATORY SCIENCE PROGRAM

Once you have decided on clinical laboratory science, an application must be processed. If applying to a program in a community college or university, admission to the institution is the first step. On admission to the college, the student may have to make a separate application to the program of interest. Admission to the college or university does not necessarily mean admission into the program.

Acceptance into the professional program in laboratory science or medical technology usually requires a minimum grade point average (e.g., 2.5, where 4.0 = A). A minimum grade point average of all science courses taken also may be stipulated. In addition, letters of recommendation, previous laboratory work experience, an interview, and a written essay may be considered in the selection process.[7]

All forms should be filled out completely and neatly. The first impression of the screening/selection committee is based on the appearance and completeness of the application materials. For example, completing forms in pencil or purple magic marker is generally not acceptable! Also, be sure to meet all deadlines.

Once the admissions committee reviews credentials, an interview is set up. The interview may be a group interview, with all members of the committee, or may be an interview with one person at a time. Other applicants may be interviewed at the same time. This interview helps to assess communication skills and maturity. Be prepared to answer likely questions, such as "Why do you want to be a clinical laboratory scientist?" It helps to think about your answer to this question before the interview. The interviewer usually allows time for you to ask questions. Show your interest and enthusiasm. It is good manners to follow-up the interview with a thank-you note for the time spent with you and indicating your continued interest. On admittance to the program, respond promptly as to whether you plan to enter the program.

REFERENCES

1. *Clinical Laboratory Science Levels of Practice.* Competence Assurance System. Washington, DC, American Society for Medical Technology, 1989.
2. *Careers in Medical Laboratory Technology.* Chicago, American Society of Clinical Pathologists, 1989.
3. *Medical Laboratory Technician: A Career for You.* Chicago, American Society of Clinical Pathologists, 1990.
4. *Medical Technology: A Career for You.* Chicago, American Society of Clinical Pathologists, 1990.
5. Board of Registry: *Procedures for Examination and Certification.* Chicago, American Society of Clinical Pathologists, 1990.
6. Gupta GC: *Allied Health Education Directory 1991*, ed 19. Chicago, American Medical Association, 1990.
7. Karni K, Oliver JS: *Opportunities in Medical Technology Careers.* Lincolnwood, Ill, NTC Publishing Group, 1991.
8. Lindberg DS, Stevenson Britt M, Fisher FW: *Williams' Introduction to the Profession of Medical Technology*, ed 4. Philadelphia, Lea & Febiger, 1984.
9. Miller-Allan PA: *Introduction to the Health Professions.* Monterey, Calif, Wadsworth, 1984.
10. Nassif JZ: *Handbook of Health Careers.* New York, Human Sciences Press, 1980.
11. NCA Examination Application. Washington, DC, National Certification Agency for Medical Laboratory Personnel, 1990.
12. *Pathology as a Career in Medicine.* Bethesda, Md, Intersociety Committee on Pathology Information, 1990.
13. US Department of Labor: *Occupational Outlook Handbook.* Scottsdale, Ariz, Associated Book Publishers, 1988.

Chapter 10

Clinical Laboratory Science Options

NUCLEAR MEDICINE

Nuclear medicine is the medical specialty that utilizes the nuclear properties of radioactive and stable nuclides for diagnostic evaluation of the anatomic or physiologic conditions of the body and provides therapy with unsealed radioactive sources. The skills of the nuclear medicine technologist complement those of the nuclear medicine physician and other professionals in the field.[5, 9, 12]

Nuclear medicine technology is concerned with the use of small amounts of radioactive materials for diagnosis, therapy, and research. Diagnosis may involve organ imaging, wherein a small amount of radioactive material (radiopharmaceutical) is administered to the patient (in vivo procedures). The radiopharmaceutical then localizes in a specific organ system where it can be detected with a special instrument called a gamma camera. Diagnosis may also involve analysis of biologic specimens, where blood or urine specimens are used to measure the level of various hormones, drugs, or other chemical substances. These are in vitro procedures, which means the radioactive pharmaceutical is added to the patient specimen outside of the patient's body. Therapeutic doses of radioactive materials can be administered to patients to treat specific diseases. Although the use of radioactive materials for therapeutic purposes is limited, research in this field promises additional applications in the future. Radioactive isotopes play an important role in developing new ways to detect tumors and other lesions and disorders. There are possibilities for using radioactive antibodies to treat certain types of tumors.

The capacity of nuclear medicine for precise and accurate measurement of tracers has made possible a variety of quite important laboratory assays in clinical diagnosis, pharmacology, biochemistry, and many other fields. The studies of blood ferrokinetics, blood volume determinations, thyroid studies, and vitamin B_{12} absorption studies for diagnosis of pernicious anemia have been of inestimable value in diagnosis and treatment. Other in vitro radioassays have sharpened diagnosis in a host of hormonal, metabolic, infectious, and inflammatory disorders.[14, 15]

Occupational Description

A nuclear medicine technologist is a highly skilled person who has studied anatomy, physiology, mathematics, chemistry, physics, basic computer programming, data analysis, radiation safety, clinical nuclear instrumentation, patient care procedures, and laboratory techniques.

Nuclear medicine technologists, under the direction of physicians who are licensed to have radioactive materials, use radionuclides and radiopharmaceuticals to perform or assist in the performance of diagnostic examinations. These examinations include radionuclidic imaging of organs and organ systems, dynamic studies, assays of body fluids and tissues, and radioassays.[14, 17]

Under supervision of a physician, the nuclear medicine technologist either directs or participates in the daily operation of the nuclear medicine department. The responsibilities are varied and can include the following:

1. Application of their knowledge of radiation physics and safety regulations to limit radiation exposure.
2. Preparation and administration of radiopharmaceuticals.
3. Operation of radiation detection devices and other kinds of laboratory equipment that measure the quantity and distribution of radionuclides deposited in the body or in a specimen.
4. Performance of in vivo and in vitro diagnostic procedures.
5. Utilization of quality control techniques as part of a quality assurance program covering all procedures and products in the laboratory.
6. Interaction with patients in a caring way.
7. Positioning patients for imaging procedures.
8. Using a computer to analyze data.
9. Participation in research and administration.[9]

Job Opportunities

The field of nuclear medicine technology is a well-developed profession and there is a constant demand for well-trained, highly skilled, and motivated nuclear medicine technologists. New developments in nuclear medicine are expected to increase the demand for technologists even further. Jobs are available in hospitals, public health institutions, government, private research institutions, the military service, commercial equipment companies, outpatient facilities, clinics, and commercial radiopharmacies. Opportunities are also available for obtaining positions in clinical research, education, and administration.

According to the 1990 Committee on Allied Health Education and Accreditation (CAHEA) Annual Supplemental Survey, entry-level salaries average $26,118.[9]

Educational Preparation

About 107 accredited nuclear medicine technology programs currently are in operation in hospitals, colleges, and universities.[9]

The technical portion of the nuclear medicine program is one year in length. Institutions offering accredited programs may provide an integrated educational sequence leading to an associate or baccalaureate degree over a period of two or four years. One-year certificate programs are also offered by some hospitals and academic institutions.[9, 12]

Prerequisites for entering nuclear medicine technology programs depend on the specific type of program offered. A strong desire to work with patients in a caring relationship is necessary. A good background in mathematics and science, coupled with a willingness to learn to use computers and technical equipment is equally important. Specific information can be obtained from the institution offering the program.

Applicants for admission must have graduated from high school or the equivalent and have acquired postsecondary competencies in human anatomy and physiology, physics, mathematics, medical terminology, oral and written communications, chemistry, and medical ethics.[9]

The curriculum includes patient care, nuclear physics, instrumentation and statistics, health physics, biochemistry, immunology, radiopharmacology, administration, radiation biology, clinical nuclear medicine, radionuclide therapy, and introduction to computer application.

Certification and Professional Organizations

In 1977, the Nuclear Medicine Technology Certification Board (NMTCB) was established as an independent certifying board. The first NMTCB examination was administered in 1978. All examinations have been designed to test the application of knowledge necessary for competent entry-level performance in nuclear medicine technology.

Those successfully passing the examination may use the title certified nuclear medicine technologist and the credential CNMT. Registrants also receive a biannual newsletter, *NMTCB News*.

Eligibility requirements include graduation from a CAHEA-accredited nuclear medicine technology program or an acceptable alternative. These alternatives include:

1. a baccalaureate or associate degree in one of the physical or biological sciences;
or
2. national certification as a registered medical technologist;
or
3. national certification as a registered radiologic technologist;
or
4. national certification as a registered nurse *and* appropriate work experience. High school graduates whose clinical experience began before Jan 1, 1987, are also eligible for the examination with six years of clinical experience.

For information about nuclear medicine technology training programs, contact:
Joint Review Committee on Educational Programs in Nuclear Medicine Technology
445 S. 300 E.
Salt Lake City, UT 84111
Telephone: (801) 355-9628

or refer to: American Medical Association: *Allied Health Education Directory*.

For additional information about the profession write or call:
　　The Society of Nuclear Medicine
　　Technologist Section
　　136 Madison Ave.
　　New York, NY 10016-6760
　　Telephone: (212) 889-0717

CYTOGENETICS

Medical cytogenetics is a highly specialized discipline dealing with the study of normal and abnormal chromosome variation. Chromosomal abnormalities are important because of their involvement in birth defects, mental retardation, and spontaneous abortion. Many genetic defects in humans can be diagnosed before birth or shortly after birth by chromosome analysis.

There are at least 3,000 genetic disorders that plague the human species. About 12 to 15 million Americans suffer from at least one of them. More than a quarter-million infants are born with physical or mental damage in the United States every year. Birth defects thus strike one out of every 14 infants born. Genetic factors are estimated to be involved in as many as 25% of disease such as diabetes, heart disease, psychiatric disorders, and some cancers.[6]

The prevalence of genetic diseases and recent technical advances in the specialized field of cytogenetics have created a critical and growing need for highly skilled cytogenetic technologists. Moreover, there are only a few clinical cytogenetics laboratories across the country that can accommodate the steadily increasing demand for diagnostic cytogenetic services. Evaluations of children with birth defects, prenatal diagnosis, and cancer studies are three major reasons for the growing emphasis on cytogenetic technology. Presently, the tremendous demand for cytogenetic services has created a need for qualified cytogenetic technologists that far exceeds the currently available number of experienced technologists by approximately 8:1.[7, 12]

Genetic screening will be in increasing demand in the future. Advances in medical genetics, including chorionic villus and amniotic fluid sampling for prenatal diagnosis, especially for women older than 35 years, have created an urgent need for skilled technologists to analyze specimens from an ever-increasing number of patients requiring these tests.

Occupational Description

Laboratory professionals who specialize in cytogenetics usually spend at least six months to a year learning how to grow cells containing chromosomes, separating and grouping chromosomes by position, and locating and identifying abnormalities. The result is a karyotype, a picture of the paired chromosomes with specific abnormalities delineated.[12]

Cytogenetic technologists are responsible for preparing specimens from bone marrow, peripheral blood, solid tissue, and amniotic fluid; performing microscopic analysis and photography of the cultured specimens; and interpreting chromosome karyotypes. Cytogenetic technologists commonly work with physicians and genetic counselors by identifying hereditary physical and mental disorders such as cystic fibrosis, Tay-Sachs disease, and Down's syndrome, to ensure that patients receive optimum treatment. Cytogenetics is becoming increasingly important in identifying hematologic disease states such as leukemias. For example, the "Philadelphia chromosome" is associated with 85% of patients with chronic myelogenous leukemia.

The allied health professionals who perform this work need patience and a keen eye to evaluate microscopic preparations, find abnormalities, and make critical diagnostic decisions based on subtle chromosomal changes.

Job Opportunities

Cytogenetic technologists work in hospital, university, research, or private cytogenetic laboratories. Currently there are over 450 of these laboratories in the United States. In addition, technologists are employed in pediatric clinics, genetic counseling clinics, research laboratories, clinical laboratories, pharmaceutical companies, and chemical industries. They may work in educational institutions in teaching or research capacities. Experienced cytogenetic technologists may be responsible for the management and operation of clinical laboratories.

Rapid development in the field of cytogenetics, especially in the last five years, has resulted in the growth of career opportunities for technologists. A job survey conducted in 1988 revealed that clinical cytogenetic laboratories have many more positions available than qualified people available to fill them. There have been more than 400 jobs available every year since 1987 with only 60 to 80 total graduates from all the formal training programs in North America available each year to fill them. Due to the high demand for, and low supply of, experienced personnel, salaries have increased significantly. Starting salaries, 1990, are generally between $20,000 to $25,000, depending on geographic location.[7]

Educational Preparation

Although cytogenetic technologists have university degrees in scientific fields, their cytogenetic training has usually been acquired on the job. In order to meet the ever-increasing demand for specialists, several university programs offering instruction and practical training in cytogenetic technology have been established since 1980. Most of these institutions have based their curricula on the statements of competence developed by the Association of Cytogenetic Technologists (ACT).

Collectively, the cytogenetic technology training programs offer a variety of educational routes into cytogenetics technology, including the awarding of a certificate, a bachelor's degree, a master's degree, or a combination of degree and certificate. All programs offering a certificate on graduation require either a degree or registration as a medical technologist prior to acceptance.[8]

Prerequisite academic coursework includes a strong emphasis on the biological and chemical sciences. Required courses generally include cell biology, microbiology, cytology or cytogenetics, genetics, general chemistry, organic chemistry, and biochemistry. Students also complete specific lecture-based and laboratory-based cytogenetics courses at the university. The clinical practicum covers all major areas of cytogenetics—blood, bone marrow, tissue, and prenatal studies.[8]

Certification and Professional Organizations

The Association of Cytogenetic Technologists, Inc. (ACT) and the National Certification Agency for Medical Laboratory Personnel (NCA) have developed a certification examination in cytogenetics. The ACT statements of competence form the basis for the questions, which are job related and criterion referenced. First administered in 1981, the examinations are now given twice yearly. More than 1,200 individuals have been certified, including 80% of the ACT membership.[8]

There are six eligibility routes available for admission to the cytogenetics examination. Eligibility is based upon educational background and work experience. The three most common routes are:

1. a baccalaureate degree with two years of full-time work experience in cytogenetics;
or
2. graduation from a structured program that culminates in a baccalaureate degree with a one-year internship in cytogenetics;
or
3. completion of a recognized cytogenetics education program (hospital or university-based) that includes a six-month internship in a full-service cytogenetics laboratory.

Examinations are also offered by the Canadian Society of Laboratory Technologists (CSLT). Approximately 150 cytogenetic technologists have been certified by the CSLT to date.[8]

The Association of Cytogenetic Technologists (ACT) is a nonprofit professional society that was established in 1975 to promote cooperation and exchange of information among those engaged in cytogenetic technology and research, and to stimulate interest in cytogenetics as a career. Membership in the association currently numbers about 1,400 technologists, laboratory supervisors, and laboratory directors working in the United States and 26 foreign countries. Although most members are technologists or supervisors, about 15% hold doctoral degrees (M.D. or Ph.D.).

The ACT sponsors an annual meeting every summer and produces four publications: a journal, a laboratory directory, a technical manual, and an information brochure. The journal, *Applied Cytogenetics* (formerly *Karyogram*), *Journal of Cytogenetics Technology*, is published bimonthly and features original research, technical and review articles, book reviews, technical tips, job placement, meeting announcements, and news.

The ACT interacts with the genetics community through its liaisons to the Council of Regional Networks for Genetic Services (CORN). The cytogenetics subcommittee of the quality assurance committee of CORN reviews cytogenetic proficiency testing programs in the United States. This subcommittee comprises representatives from the ten regional genetics networks, state health departments, the federal government, and ACT.

The American Society of Human Genetics (ASHG) provides educational and scientific programs for students interested in genetics. Students are eligible for membership in ASHG as well as the ACT.

For information on certification examinations, contact:
The National Certification Agency for Medical Laboratory Personnel (NCA)
2021 L St. N.W., Suite 400
Washington, D.C. 20036
Telephone: (202) 857–1023

For information on membership in the Association of Cytogenetic Technologists, Inc. (ACT), contact:

>Helen Bixenman
>Membership Secretary
>Genetrix
>6401 E. Thomas Rd.
>Scottsdale, AZ 85251
>Telephone: (602) 945-4363

For additional information on ACT, contact:

>Mike Zvarich
>Public Relations Officer
>Genetics Associates of North Carolina
>120 Conner Dr., Suite 201
>Chapel Hill, NC 27514
>Telephone: (919) 942-0021

CYTOGENETIC TECHNOLOGY EDUCATIONAL PROGRAMS: OCTOBER, 1989

1. Post Diploma Specialty Training Program in Cytogenetics

>Toronto Institute of Medical Technology
>222 St. Patrick St.
>Toronto, Ontario, Canada M5T 1V4
>(416) 596-3112

2. School of Cytogenetic Technology

>LAC-USC Medical Center
>Women's Hospital, Room 1-M 20
>1240 N. Mission Rd.
>Los Angeles, CA 90033
>(213) 226-3007

3. Human Cytogenetic Technology Certificate Program

>Department of Biology
>California State University, Dominguez Hills
>1000 E. Victoria St.
>Carson, CA 90747
>(213) 516-3381

4. Medical Cytogenetic Technology Program

School of Allied Health Professions
The University of Connecticut
358 Mansfield Rd., U-101
Storrs, CT 06169-2101
(203) 486-0036

5. Program in Cytogenetics

School of Allied Health Sciences
The University of Texas Health Science Center at Houston
P.O. Box 20708
Houston, TX 77225
(713) 792-6330

6. Cytogenetics Laboratory Technology

British Columbia Institute of Technology
3700 Willingdon Ave.
Burnaby, B.C., Canada V5G 3H2
(604) 432-8296

7. Biological Sciences: Option in Cytogenetic Technology

School of Science and Allied Health
Department of Biology
Kennesaw State College
Marietta, GA 30061
(404) 423-6158

8. Cytotechnology and Cytogenetics

College of Allied Health Sciences
Edison Bldg., Room 1924
Thomas Jefferson University
130 S. 9th St.
Philadelphia, PA 19107
(215) 662-3233

9. Cytogenetic Technology

Medical Technology Education Dept.
Medical University of South Carolina
171 Ashley Ave.
Charleston, SC 29425-0701
(803) 792-3169

10. Cytogenetic Technologist's Training Program

W101GH
University of Iowa Hospital and Clinic
Iowa City, IA 52242
(319) 356-3877

11. Cytogenetics Program

Department of Biology
Hofstra University
Hempstead, NY 11550
(516) 463-5527

12. Cytogenetics Technologist Certification Program

Cytogenetics Laboratory
Department of Pathology
Mercy Hospital and Medical Center
Stevenson Expressway at King Drive
Chicago, IL 60616-2477
(312) 567-2283

CYTOTECHNOLOGY

Cytology is the study of the structure and the function of cells. The function of a cytopathology laboratory is ultimately that of cancer diagnosis. This is done primarily with the microscope used to screen slide preparations of body cells for abnormalities, indicating either benign or malignant conditions. Traditionally, this is often a cancer diagnosis made earlier than is possible by other methods, while cure rates are still high. The cervical/vaginal "Pap" smear is the most common cell sample submitted to the cytology laboratory. Routine use of the Papanicolaou test has been directly responsible for reducing cancer of the cervix from the number one cause of cancer death in women in the 1940s to number six in the 1980s. Cell specimens may be obtained from other various body sites, such as the oral cavity, the lung, the breast, or any body cavity that sheds cells.[9]

Occupational Description

A cytotechnologist is the highly specialized laboratory professional trained to detect abnormal cells and to provide the preliminary diagnostic interpretation on cellular samples. The technologist performs these tests by carefully examining all of the cells on each microscope slide for subtle differences in size, shape, arrangement, and color.

If the sample is a routine gynecologic (Papanicolaou) smear and it is interpreted as being "normal" or "negative for cancer cells," the cytotechnologist (CT) can issue a final report. If cells appear to be abnormal, or if it is from a nongynecologic source (such as sputum, urine, body cavity fluid, or liver aspirate), significant cells are marked and the slide is passed on to the cytopathologist along with the cytotechnologist's diagnostic impression. Final reports on these cases are the responsibility of the cytopathologist.

Although cytotechnologists work primarily in the area of cancer detection, they are also trained to use a variety of techniques to detect other abnormal conditions, such as hormone imbalance, that are evidenced by cellular structure in the body. In recent years, fine needles have been used to aspirate lesions, often deeply seated in the body; this greatly enhances the ability to diagnose tumors located in inaccessible sites.[9, 16]

Cytotechnologists may be directly involved in specimen procurement. They may obtain buccal smears and assist clinicians with needle biopsy aspirations. In the latter role, they make sure the specimen is adequate and properly preserved.[10]

A few years ago, cytotechnologists were expected merely to indicate a general diagnosis of "cancer." However, now the technologist performs special stains to help define the malignancy further.

Cytotechnologists work independently with little supervision. They must be patient, precise, and have relatively good eyesight. The work should be of particular interest to the physically handicapped, especially those who have difficulty standing. Above all, cytotechnologists must enjoy making decisions and taking responsibility, since their findings will directly affect a patient's course of treatment.[1, 13]

The future promises to bring a computer that would scan cells samples, identify abnormalities, and record the images that are not normal for later viewing by the cytotechnologist. Development of this computer is particularly timely as the demand for cytotechnologists increases in the 1990s.

Job Opportunities

Today, there are more jobs for cytotechnologists than educated people to fill those jobs. The future long-term employment looks bright—well into the next century. The need is great throughout the country.[1]

In 1990, the national average beginning salary for cytotechnologists was approximately $27,772, although salaries vary by area of the country.[1, 9]

Most cytotechnologists work in hospitals or in private laboratories, while some prefer to work on research projects or teach. Cytotechnologists have an unlimited choice of practice settings. Hospitals, for-profit laboratories, clinics, nursing homes, public health facilities, business, and industry currently have positions open for qualified cytotechnologists. Opportunities in management also exist, particularly if the cytotechnologist has advanced education.[1, 9]

Educational Preparation

A combination of formal education leading to a Bachelor of Science degree, plus clinical education in a cytotechnology (CT) program accredited by the Committee on Allied Health Education and Accreditation (CAHEA) is required for a career as a cytotechnologist. Most cytotechnologists complete three years of academic coursework. Their fourth year is spent in a 12-month hospital or laboratory internship. Upon completion of the internship, a Bachelor of Science degree is granted (3 + 1 program). A few university-based integrated (2 + 2) programs are available as well as the 4 + 1 option, in which students complete their B.S. degree prior to the internship.[1]

To prepare for a career as a cytotechnologist, a solid foundation in high school sciences is needed—biology, chemistry, mathematics, and computer science.[1]

College coursework should include successful completion of at least 20 semester hours (30 quarter hours) in the biological sciences, chemistry courses equaling or exceeding eight semester hours (12 quarter hours), and some mathematics.[9]

The cytotechnology curriculum includes the historical background of cytology, cytology as applied in clinical medicine, cytology in the screening of exfoliate tumor cells, and areas of anatomy, histology, embryology, cytochemistry, cytophysiology, endocrinology, and inflammatory diseases.[9, 12]

Certification and Professional Organizations

The Board of Registry of the American Society of Clinical Pathologists gives a national certification examination. Students take this examination after meeting their academic and laboratory education requirements. Those who pass the exam for cytotechnology may use the initials, CT(ASCP), after their name.[1]

A cytotechnologist with a bachelor's degree and five years' experience, or a master's degree and four years' experience, or a doctoral degree and three years' experience can qualify to be a specialist in cytotechnology, SCT(ASCP). These specialists are skilled in examining all types of body specimens, including needle aspirates and fine needle aspirates.[1]

The American Society for Cytotechnologists (ASCT) is run by cytotechnologists. The ASCT keeps members informed about current issues and legislation. Student membership is available.

The International Academy of Cytology (IAC) has a registry examination and makes continuing education available to its members. The IAC serves as the only registry examination in many countries that do not have their own national certifying agencies. It also publishes the journal, *Acta Cytologica*.

State societies have also been established. These societies generally offer regional and local meetings, lectures, workshops, etc. Membership for students is free.

Information on careers, curriculum, and certification/registration may be addressed to:

American Society of Cytology
1015 Chesnut St., Suite 1518
Philadelphia, PA 19107
(215) 922–3880

American Society of Clinical Pathologists
Board of Registry
P.O. Box 12270
Chicago, IL 60612
(312) 738-1336

HISTOLOGY

Histology is a structural science concerned with studies of cellular morphology, chemical composition, and function of normal and abnormal tissue.[9]

The birth of histologic technique dates back to 1664 when Robert Hooke cut sections of cork with his pen knife and observed them under a microscope. In 1670, van Leeuwenhoek made sections from a writing feather, a bovine optic nerve, and the centers of dried flowers by using his hand-sharpened shaving razor.[3]

Cancer can often be detected by the appearance of cells in a tissue sample. Once a sample of tissue is taken from the patient, it is sent to the laboratory. It is the job of the histology technician to prepare very thin sections of the tissue sample for microscopic examination by a pathologist.[2]

Occupational Description

Histotechnology is a varied career with many specializations. The main responsibility of the histology technician/technologist (HT/HTL) in the clinical laboratory is preparing sections of body tissue for examination by a pathologist. This includes the prepartion of tissue specimens of human and animal origin for diagnostic, research, or teaching purposes. Tissue sections prepared by the histologic technician/technologist for a variety of disease entities enable the pathologist to diagnose body dysfunction and malignancy.[9]

The histology technician must work quickly and under pressure, since the answers may be needed while the patient is in surgery. The frozen section technique assists surgeons during surgery. Suspicious tissue is sent from the operating room to the histotechnologist. Working closely with the pathologist, the histology technician freezes and cuts the tissue specimens, mounts them on slides, and stains them with special dyes to make details of the cell visible under the microscope. With the information learned from the section of tissue biopsy, the pathologist and surgeon determine if disease is present and if it has spread. They can then decide on the best course of treatment for the patient.[2, 3, 11]

Histotechnicians process sections of body tissue by fixation, dehydration, embedding, sectioning, decalcification, microincineration, mounting, and routine and special staining. After fixing the tissue, technicians put it on a tissue processor, which removes the fluids, making the specimen clear or transparent; then the specimen is soaked with paraffin and put on a special cutting tool called a microtome. This instrument can cut the section of tissue into very thin slices ($1/5,000$ in.). Technicians then put the thin sections or section on a microscopic slide, stain it, and cover it with a very thin glass. The pathologists look at this specimen under a microscope.[9, 11]

Histology technicians can do special stains on microscopic slides to show fungus, bacteria, connective tissue, and other disease activity. They may process specimens and treat them with plastics rather than paraffin so that they can cut the specimen much thinner (1/25,000 in.). They can also make slides with cross-sections of bone. Some histology technicians work with electron microscopy.[11]

A histology technician (HT) who earns a baccalaureate degree and either has one year of experience or attends a CAHEA accredited histotechnology program can become a histotechnologist (HTL). Histotechnologists perform all the functions of the histotechnician as well as the more complex procedures for processing tissues. They identify tissue structures, cell components, and cell staining characteristics, and relate them to physiological functions. Histotechnologists implement and test new techniques and procedures, make judgments concerning the results of quality control measures, and institute proper procedures to maintain accuracy and precision. Histotechnologists perform such complex techniques as enzyme histochemistry, immunohistochemistry, and electron microscopy. A histotechnologist can also teach, be a supervisor in the laboratory, or be the director of a school for histologic technology.[2, 9]

All histology technicians have certain common characteristics: they have an interest in science; they are problem-solvers; they like challenge and responsibility; they are detail-oriented, precise, accurate, reliable, emotionally stable, work well under pressure, and are able to finish a task once started; they communicate well, both verbally and in writing; and they like work in which they deal with "things". Skillful hands, good eyesight, and normal color vision are also important.[2, 11]

Job Opportunities

Today, there are more jobs for laboratory personnel than educated people to fill those jobs. The future long-term employment looks bright—well into the next century. The need is great throughout the country.[2]

Most histology technicians/technologists work in hospital laboratories, averaging a 40-hour week. According to the 1990 CAHEA Annual Supplemental Survey, entry-level salaries average $19,339 for histotechnicians, although salaries vary by area of the country. The national average beginning salary for histology technologists was approximately $24,000 in 1989.[9, 12]

Histology technicians have an unlimited choice of practice settings; hospitals, for-profit laboratories, clinics, nursing homes, public health facilities, business, and industry currently have positions open for qualified histologic technicians. Other opportunities for histology technicians are in industrial research, veterinary pathology, marine biology, and forensic pathology.[2]

Educational Preparation

Training for histology technicians began in 1936. It was not until the 1960s, however, that the American Medical Association (AMA) set guidelines for programs of study.[11]

To prepare for a career as a histology technician a solid foundation in high school sciences is important: biology, chemistry, mathematics, and computer science. Clinical education in a histologic technology (HT) program accredited by CAHEA is also required.[2]

To be a histology technician (HT) requires a high school diploma or an associate degree. Histologic technology programs can be found in community colleges that offer an associate degree or in hospitals that offer certificate programs.[2]

Currently, the National Accrediting Agency for Clinical Laboratory Sciences (NAACLS) approves the schools for histology training. At this time there are fewer than 40 active schools.[9, 11]

Most hospital-based histotechnician (HT) certificate programs are 12 months long, unless the curriculum is an integral part of a college program. For the histotechnologist (HTL), a baccalaureate degree program of four years is required.[9]

The curriculum includes both didactic instruction and practical demonstration in the areas of medical ethics, medical terminology, chemistry, laboratory mathematics, anatomy, histology, histochemistry, quality control, instrumentation, microscopy, processing techniques, preparation of museum specimens, and record and administration procedures. The baccalaureate level program includes coursework designed to prepare supervisors and teachers with advanced capabilities.[9, 12]

It has been recommended that the curriculum be an integral part of a junior or community college program culminating in an associate degree, and the course of study include chemistry, biology, and mathematics. Six months of the program must be dedicated to clinical training in an associate degree program.[9, 10, 12]

Certification and Professional Organizations

The Board of Registry of the American Society of Clinical Pathologists gives a national certification examination, which consists of two parts: a practical portion and a written portion for both the technician and the technologist. After they pass the examination, technicians may use the initials HT(ASCP) after their names; technologists use the initials HTL(ASCP). In a few states, HTs must be licensed.[2, 11]

To be eligible to take the HT(ASCP) examination, one of the following must be documented:

1. Successful completion of a CAHEA accredited histotechnology program;

or

2. Associate degree or at least 60 semester hours of academic credit with 6 semester (9 quarter) hours in chemistry and 6 semester (9 quarter) hours in biology and 1 year full-time acceptable experience in histopathology;

or

3. High school graduation or equivalent and 2 years full-time acceptable experience in histopathology.[4]

Eligibility for the histotechnologist HTL(ASCP) examination may be met by one of the following:

1. Baccalaureate degree from a regionally accredited college/university with 12 semester (18 quarter) hours in chemistry, 16 semester (24 quarter) hours in biology, selected from general biology, histology, zoology, anatomy, or physiology; 4 additional semester (6 quarter)

hours in other unspecified science courses; 3 semester (4 quarter) hours in mathematics, *and* one-year full-time acceptable experience in a histopathology laboratory;

or

2. Baccalaureate degree from a regionally accredited college/university including the above course requirements *and* successful completion of a CAHEA accredited histotechnology program.[4]

The National Society for Histotechnology (NSH) was incorporated on July 22, 1974, in the state of Virginia, as an organization dedicated to the study and practice of histotechnology.

Benefits of NSH Membership include the following:

1. Information, communication, and networking: The Society publishes a journal entitled, *Journal of Histotechnology*, and a newsletter, *NSH in Action;*
2. Educational opportunities—meetings, extension seminars, educational aids, training aids, educational resources, and teleconferences;
3. Quality control records and hematoxylin and eosin (H&E) guidelines;
4. Scholarships and awards for members and students;
5. Additional benefits—liability insurance policy, employment opportunities updates, promotional material, and vocational/career information.

For further information on accredited programs, the reader is referred to the American Medical Association's *Allied Health Education Directory.*

For further information on careers/curriculum and certification/registration, contact:

National Society for Histotechnology
5900 Princess Garden Pkwy.
Suite 805
Lanham MD 20706
(301) 577-4907
American Society of Clinical Pathologists
Board of Registry
Box 12270
Chicago, Il. 60612
(312) 738-1336

Acknowledgment

Special thanks to Susan Dingler, CT(ASCP), for her contributions to the cytotechnology section of this chapter.

REFERENCES

1. American Society of Clinical Pathologists: *Cytotechnologist: A Career for You.* Chicago, ASCP, 1990.
2. American Society of Clinical Pathologists: *Histologic Technician: A Career for You.* Chicago, ASCP, 1990.
3. *The Art and Science of Histotechnology.* Miles Inc, Elkhart, Ind, 1987.

4. Board of Registry: *Procedures for Examination and Certification.* Chicago, American Society of Clinical Pathologists, 1990.
5. Division of Allied Health Education and Accreditation: *Essentials and Guidelines of an Accredited Educational Program for the Nuclear Medicine Technologist.* Chicago, American Medical Association, 1984.
6. *Facts: March of Dimes Preventing Birth Defects.* White Plains, NY, March of Dimes Birth Defects Foundation, p 2.
7. Gasparini RP: Human resource data and national salary statistics for cytogenetic laboratories throughout the United States. *Karyogram* 1988; 14:102–111.
8. Gasparini RP, Kaplan BJ, Stevens LR: Innovations in human genetics education *Am J Hum Genet* 1988; 42:200–203.
9. Gupta GC: *Allied Health Directory 1991*, ed 19. Chicago, American Medical Association, 1991.
10. Heustis DG: New visibility in the lab for cytotechnologists. *MLO* 1985; 17:64–70.
11. *Histologic Technicians*, brief 227. Chronicle Guidance Publications, Moravia, NY, 1988, pp 10–12.
12. Karni K, Oliver JS: *Opportunities in Medical Technology Careers.* Lincolnwood, Ill, NTC Publishing Group, 1990.
13. Lindberg DS, Britt MS, Fisher FW: *Williams' Introduction to the Profession of Medical Technology*, ed 4. Philadelphia, Lea & Febiger, 1984.
14. Miller Allan P: *Introduction to the Health Professions.* Monterey, Calif, Wadsworth, 1984.
15. Myers WG, Wagner HN: Nuclear medicine: How it began. *Nucl Med* (reprint 1975);
16. Nassif JZ: *Handbook of Health Careers: A Guide to Employment Opportunities.* New York, Human Sciences Press, 1980.
17. Position description: Nuclear medicine technologist. *J Nucl Med Tech* 1979; 7:178–181.

BROCHURES AND PRINTED MATERIAL AVAILABLE FROM:

Association of Cytogenetic Technologists, Inc. (ACT)
The National Society for Histotechnology (NSH)
Nuclear Medicine Technology Certification Board (NMTCB), The Society of Nuclear Medicine Technologist Section
Thomas Jefferson University Catalog, 1989–91
School of Allied Health Science Catalog 1989–91, The University of Texas Health Science Center at Houston
The National Certification Agency for Medical Laboratory Personnel (NCA)

Chapter 11

Employment and Career Opportunities

Clinical laboratory scientists are found in many settings—almost everywhere that professional health services are provided. Opportunities exist here and abroad and in traditional settings—such as hospitals—and nontraditional settings—such as NASA.[5]

An undergraduate degree in clinical laboratory science (CLS) is one of the most versatile degrees offered in higher education. Table 11-1 summarizes employment and career opportunities, but it is certainly not complete. These career paths should not be considered "alternate careers," since the word "alternate" implies some field other than clinical laboratory science. Every course taken by clinical laboratory science students prepares them for opportunities outside the traditional clinical laboratory setting. The list grows daily as technology develops and the creativity of laboratory professionals expands.

Solberg and Karni, in a 1987 survey of clinical laboratory science graduates from the University of Minnesota,[8] suggest that once clinical laboratory scientists have graduated they tend to remain within CLS or related fields and represent a stable work force. Almost 80% held a laboratory-related position in a present or past job.[8] These findings were confirmed by Gregg and others in a survey of Michigan Technological University CLS graduates. They found that of all positions CLS graduates held since graduation (total, 1,086), 648 (59.7%) were in hospital clinical laboratories. These positions were mostly technical: generalist positions (82%) or positions in a single department (24.9%).[4, 5]

Your first occupational decision is to choose a career course. You might be one of the many laboratory professionals who decide to stay put. There's nothing wrong with that as long as it is your choice.

Your career goals while attending college may change or develop as you mature. Your personality, strengths, and weaknesses will change slightly as you age. What once was an exciting, challenging job may someday seem monotonous and routine. A degree in clinical laboratory science gives you the flexibility to make changes!

There are a few major questions you must answer:

TABLE 11–1.
Career and Employment Opportunities

I. Positions available in health care
   A. Laboratory—technical positions
      1. Hospitals
      2. Private laboratories
      3. Public health laboratories
      4. Physicians' office laboratories (POLs)
      5. Health maintenance organizations (HMOs)
      6. Government—VA hospitals
      7. Crime laboratories, forensics
      8. Research and development
      9. In vitro fertilization laboratories
      10. Pathology assistant
      11. Flow cytometry and transplant services
      12. Blood banks, American Red Cross
      13. Environmental laboratories
      14. Veterinary medicine
      15. Humanitarian agencies—Project Hope, Peace Corps
   B. Nontechnical, health care–related positions
      1. Inspectors—accreditation agencies such as JCAHO and CAP
      2. Infection control officer, epidemiology
      3. Utilization review coordinator
      4. Quality assurance director
      5. DRG coordinator
      6. In service education coordinator
      7. Human resources director—personnel
      8. Management—laboratory director, controller, administrative vice president
      9. Director of safety, waste control manager
      10. Director of marketing
      11. Information specialist
      12. Planning and development
      13. Private consulting
      14. Office laboratory management
II. Outside health care
   A. Education
      1. Teaching—university, community college, hospital, medical center
      2. Administrative positions—department head, dean, higher education administrator
   B. Research
      1. University, industrial, pharmaceutical
   C. Management positions, administration and supervision
   D. Industrial and Commercial Positions
      1. Research
      2. Sales
      3. Marketing
      4. Management
      5. Employee health service
      6. Technical writing, medical illustrations
      7. Technical field representative
      8. Product manager
      9. Occupational health and safety officer

                10. Insurance underwriter or risk management officer
                11. Chemical and natural resource companies
            E. Military, government
                1. Officer positions
                2. Biometrist
                3. Agency positions—Centers for Disease Control (CDC), Health and Human Services (HH&S)
                4. NASA
    III. Careers requiring further education
        A. Graduate school
        B. Medical, dental, and veterinary school
        C. Education

1. Do you want a technical position? What settings attract you?
2. Do you want a position in health care, but nontechnical?
3. Do you want a position outside of health care?

Answering these questions will help you narrow your career choices.

The career paths of most clinical laboratory scientists usually include a few years of laboratory bench experience. Some then change to positions outside of the clinical laboratory and outside of health care. A number of these career paths require additional education and work experience.

It is important to know yourself and to know what is important to you in a career. Do you like working with people and/or patients? Do you prefer to work with instruments, chemicals, etc.? Do you wish to work independently? Do you require a lot of freedom to structure your own job? Are you a self-starter? Do you like automation and trouble-shooting? Are opportunities for advancement important to you? Do you have good teaching ability? Are you willing to alter your work schedule, including acceptance of night and weekend assignments? Do you like to travel?

What do you particularly enjoy? When you think about career opportunities, consider the body of knowledge you already have. What content areas interest you? What skills have you developed? Most laboratory professionals are good problem-solvers, logical, good organizers, and good planners. These skills are essential to most positions described in Table 11-1.

With these questions in mind, you can explore the career opportunities described in Table 11-1; this chapter is organized in the same sequence.

LABORATORY TECHNICAL POSITIONS

The greatest number of clinical laboratory scientists are presently employed in hospitals. The working conditions, salary, and opportunities vary with the size and type of hospital.

It is predicted that clinical laboratory scientists will assume more supervisory and managerial responsibilities in a clinical laboratory setting. They will orient and train less-qualified employees, and share duties in safety, quality control, and inventory control.[9]

Hospital

The small hospital (fewer than 100 beds) offers many challenges to the laboratory professional. A pathologist may or may not be on the medical staff, full-time. The laboratory staff usually function as generalists. They are often responsible for specimen collection, including venipuncture. A small hospital laboratory may not have a permanent afternoon or midnight shift. Laboratory personnel are often required to share "on-call" staffing on weekends, afternoons, midnights, and holidays. Emergencies happen all the time! The clinical laboratory scientist is usually reimbursed for being available as well as for actual time spent performing laboratory tests. A very small hospital also may expect laboratory personnel to take x-rays or perform ECGs.[6]

Individuals wishing to do a variety of laboratory tests in a variety of departments may find the small laboratory environment satisfying. The laboratory professional will function independently and may find the working relationships with the other hospital staff rewarding. Wages may be comparable to or slightly lower than provided by larger hospitals. However, the towns in which these smaller hospitals are located may have a lower cost of living. The potential for earning more money also depends on the amount of overtime.

Life in a small town has its trade-offs. Social and cultural activities may be limited, but the crime rate and traffic problems may be less. Vacations and time off may be difficult to schedule. The laboratory itself, because of lowest volume and limited space, is not likely to have the most up-to-date equipment. More manual laboratory tests will be performed. Many of the more sophisticated laboratory tests will be sent to a reference laboratory rather than be performed on site.[6]

A medium-sized hospital (up to 300 beds) will have a laboratory that is somewhat departmentalized. The hospital usually has at least one pathologist on site and an administrative technologist or laboratory manager. Most of these hospitals will have round-the-clock coverage, with at least an afternoon shift. The technologists may rotate through various departments or may remain in one department. Rotating technologists are an asset to the laboratory. They compensate for shortages in various departments when illness or vacations occur.

The sophistication and automation of laboratory instrumentation is generally determined by the test volume. Some tests will be referred to a larger laboratory. Clinical laboratory scientists work weekends, and perhaps afternoon and midnight shifts, as part of their assignment.[6]

A large hospital (more than 300 beds) will have an organization chart that includes pathologists, laboratory managers, department heads, and/or section chiefs. The laboratory manager or chief technologist may have a variety of duties, including personnel hiring and evaluation, scheduling, or equipment and supply purchasing. A department head or section chief is generally responsible for one department or specialty area. He or she may be responsible for interviewing and hiring employees, developing new test procedures, supply ordering, and teaching.

Work schedules may be very innovative and not necessarily a 7:00 to 3:30 shift five days a week. The fringe benefits, such as retirement, life insurance, and health insurance, are generally competitive with industry standards. When considering a position, realize these benefits could be worth 20% to 33% of your salary.

Workers may feel somewhat isolated in a large organization. The chain of command is more complicated and communication problems increase with the size of the laboratory. Opportunities for advancement to department head or manager may be limited and slow. Most clinical laboratory scientists are specialized in large laboratories, allowing them to study one area in great detail. Their knowledge and skills in the other areas, though, may become rusty after awhile.[5,6]

Private Laboratory

Independent laboratories and hospital laboratories differ in focus. Hospital laboratories exist first and foremost to serve the needs of the inpatient, while independent laboratories focus on the outpatient market.

The size and type of private laboratory varies. It may range from a very small physicians' office laboratory (POL), performing minimal testing, to a large reference, commercial laboratory.

One may have considerable autonomy in a small physicians' office laboratory, being "the boss." There are usually not night calls, but the laboratory personnel may work weekends if the physicians have scheduled office hours. Some POLs offer only six or eight of the most common tests and may be the size of an apartment kitchen. Large group practices, on the other hand, may have a laboratory that rivals a small hospital facility. POLs will be regulated under the Clinical Laboratory Improvement Amendments of 1988. Laboratory personnel requirements in this legislation will provide more opportunities for clinical laboratory scientists as bench technologists and as private consultants, assuring the POLs comply with these new regulations. Consultants may train personnel, establish and monitor quality control procedures, and purchase supplies and instruments. Management consultant firms will also hire clinical laboratory scientists with laboratory experience and advanced degrees (such as an M.B.A.).[5-7]

Health maintenance organizations (HMOs) employ laboratory personnel as staff members, supervisors, and administrators.

Commercial and Industrial Laboratories

There are diverse career opportunities in large reference laboratories. It is often not feasible for a small-volume clinical laboratory to perform all of the specialized esoteric tests. Reference laboratories perform these tests and are usually part of a larger corporation, such as Smith-Kline Bio-Science, and Hoffman-LaRoche.

Laboratory personnel may be employed as technical specialists, actually performing the laboratory work. They may work in "research and development," developing and evaluating new tests, methods, and instruments. Laboratory personnel may function as educational consultants, doing in-service training and offering workshops, seminars, and continuing education to clients.

The salaries are comparable to those of laboratory workers in larger hospitals. Benefits are good, and such options as stock-buying and tuition reimbursement are often available. Laboratory personnel may receive "bonuses" and "incentives" in larger commercial laboratories, based on profits made by the company. Sometimes, however, the laboratory worker feels like a small cog in a big machine. The company may have a production line orientation. There is emphasis on efficiency, speed, and quality. There is limited patient contact and limited contact with other health professionals. Opportunities for advancement are available, particularly with additional education.[5, 6]

Public Health Laboratories

Public health laboratories may be under city, county, or state control. Most positions are under civil service control. The type of laboratory testing might be limited to serology and microbiology testing, or may be more diverse. Laboratory tests include the identification of organisms causing outbreaks of infectious and communicable diseases. Salaries are usually competitive. There is little weekend work.

Veterans Administration and Other Federal Hospitals

The patient population and the types of tests performed are limited in most federal hospitals. There are no maternity or pediatric wards in most VA hospitals, for example. Laboratory personnel in VA hospitals are civil service employees. They are paid under a general schedule (GS), and the pay is generally competitive with other hospitals and independent laboratories. Fringe benefits are excellent. Advancement is possible when openings develop and when one qualifies through work experience or by passing a civil service examination. Preferential treatment is often given to individuals already in the "system" as advancement opportunities arise.

There is a certain amount of paperwork and bureaucracy in all federal institutions. This may require that laboratory personnel punch time clocks. Administrative policies may appear to be rigid and complex, but there is employment security in most civil service positions.[5, 6]

Other Sites for Laboratory Bench Positions

County, state, and national crime or forensic laboratories employ laboratorians to analyze body fluids, such as blood or semen, as well as other substances from crime-related deaths or injuries. Testing may include blood typing, hair analysis, poison studies, histocompatibility testing, ballistic studies, and alcohol and drug studies. The FBI, CIA, and private laboratories have positions available for clinical laboratory scientists who enjoy detective work.[5-7]

In vitro fertilization laboratories provide opportunities for clinical laboratory scientists. These laboratories assist in the attempt to help infertile couples bear children, often by fertilization of the ovum outside the woman's body and implantation of the embryo into the woman's uterus. Clinical laboratory scientists may coordinate the sperm bank and perform assays required for the artificial insemination laboratory.[5, 7]

Blood banks employ a number of laboratory personnel. In these organizations, such as the American Red Cross, blood is drawn, typed, and checked for any unexpected antibodies, as well as for infectious agents such as hepatitis or the AIDS virus. There is little patient contact.

Flow cytometry is a new technology that uses cell surface markers and ploidy determinations on specimens such as blood cells. Technologists examine specimens from patients with immune deficiencies, leukemias, and other malignancies.[7]

Opportunities are increasing in transplantation laboratories and clinics with a rise in the number of kidney, heart, liver, and bone marrow transplants. Tissue and cell type matches, reconstitution, and specimen preservation are required.[7]

One of the most recent trends has been for clinical laboratory scientists to find positions in environmental laboratories. They may test soil, air, milk, sewage, and botanical specimens utilizing laboratory techniques similar to those learned during their education and training.[7]

Veterinary medicine offers numerous opportunities in research, in the veterinarian's office, and in zoos. The collection of blood from animals is an interesting experience when one considers drawing blood from the tail of a mouse or the ear of a rabbit. Normal laboratory test values of animal blood are different from those of human blood.

Many positions are available for laboratory personnel and health educators overseas, particularly in the Middle East countries. These positions may or may not be associated with a humanitarian interest, such as the Peace Corps, Project Hope, and the World Health Organization (WHO). Responsibilities may include training other health care personnel, creating schools, dispelling local superstitions, and representing our country. Because living conditions are poor in most cases, the candidate should be determined to adapt to a variety of situations.[5-7]

Nontechnical Health Care–Related Professions

There are positions available with laboratory accreditation agencies, such as the Joint Commission on the Accreditation of Healthcare Organizations (JCAHO) and the College of American Pathologists (CAP). Both groups inspect laboratories for accreditation purposes. Inspectors visit the laboratory to make sure the laboratory is meeting certain standards, such as maintaining current procedure manuals and maintaining a quality control/quality assurance program.[5]

Infection control, the field of epidemiology, is concerned with keeping the institution free of infectious agents. Investigation is made of nosocomial (hospital-acquired) infections and studies of areas that have developed a high incidence of certain infections. Nurses often function in this capacity in hospitals, but clinical laboratory scientists with a strong interest in microbiology and epidemiology are equally qualified. The career involves detective work, diplomacy, and teaching skills.[5, 7]

Utilization review coordinators review patient records to make sure the correct diagnostic tests were ordered, that treatment was necessary, and the medical care was appropriate. With the implementation of diagnosis-related groups (DRGs) as the federal method of cost reimbursement, this function has become very important.

Certain laboratory personnel may be designated to plan, implement, and oversee quality assurance (QA) programs for the laboratory and the institution. Standards for quality care are determined and corrective action is taken as appropriate. Within the laboratory, QA involves the entire process of patient specimen testing, from the collection of the specimen, performing the laboratory test using reliable methodology and appropriate controls, to reporting the test in a timely and accurate manner. Quality assurance includes a structured protocol for instrument maintenance and calibration and use of reagents, controls, standards, and supplies that are effective and accurate. Quality assurance programs also include evaluating personnel and providing continuing education.

The QA director develops reports for the hospital's administration regarding the accuracy and reliability of the clinical laboratory results. The director works with hospital staff and government inspectors in setting criteria against which laboratory results are compared.

DRG coordinators are necessary to help the laboratory comply with federal standards. This area may be presently underserviced by clinical laboratory scientists.[7]

There are positions in health care for safety directors and waste control managers. All industries, including hospitals, are under close surveillance for their waste control and disposal methods. There is particular concern about hazardous wastes such as needles, syringes, specimens, and blood products.[5,7]

Health care personnel placement specialists are needed in medical centers, employment agencies, and universities. The specialist must have a knowledge of government regulations and must have good interviewing skills.[7]

Many laboratories have computers and use laboratory specialists to write programs, "debug" package programs, and provide the necessary computer interfaces among the laboratory, patient records, and other health professionals. On the management level, laboratories need help in developing sophisticated information systems, retrieving data, utilizing software, and storing and analyzing data.[5,7]

OUTSIDE HEALTH CARE

Education

Education of clinical laboratory personnel, at the bachelor of science and associate degree levels, takes place in a variety of settings. In most states, 3 + 1 hospital-based programs educate many of our CLS and MT students. University-based programs, with a variety of formats such as 2 + 2 exist in all states and vary in number. Most CLT and MLT students are taught at the junior college or community college level.

All of these types and levels of professional education require on-site clinical instructors, in addition to education coordinators and program directors. Even though a position in a hospital may primarily be a technical position, instruction of students and new employees may be required.

When considering a career in academia, first consider the various institutions of higher education that are available. The two-year colleges, primarily community colleges, tend to emphasize quality teaching and college service, such as serving on college committees. At four-year colleges and universities there tends to be emphasis on research as well. At research universities and graduate schools research is required.

Future academicians must also determine whether an institution has a tenure track, and, if so, what the requirements are for receiving tenure. Tenure is generally granted after six years of teaching. The difficulty that most CLS educators face in the tenure process is establishing a research record. If you wish to pursue a position at a major university, seriously consider entering and completing a doctoral program. Begin to collaborate with other researchers early in your career.[3]

Autonomy is very important to most academicians. That means when you're not formally in a classroom, you decide when you arrive and when you leave; you decide what your work week is; you decide how many manuscripts you will write and how many presentations fit into your plans.[3]

Those particularly suited for an academic career generally enjoy working with students and teaching. They are curious and enjoy learning. They are able to clearly explain concepts and arrange what they want to say in an organized presentation.

Faculty positions at universities and colleges generally require a master's degree. There may be some faculty positions available at community colleges, where a bachelor's degree background is satisfactory.

Graduate education is mandatory for most academicians to be promoted and tenured. Advancement to higher administrative positions almost always requires a Ph.D or an Ed.D. degree.

There is a certain amount of prestige in being a college or university professor. The fringe benefits are excellent and usually include tuition reimbursement for yourself and family members, good retirement benefits, health care, and life insurance. Long vacations and sabbatical leaves are other benefits. Advancement is possible through the promotion system. One may be hired as an instructor and advance to full professor. Salaries are dependent on the geographic location, the institution, the market, etc. If you consider that it may be a nine-month appointment, the salary looks quite good.

For those not interested in research or graduate education past the master's degree level, a position in a hospital school of medical technology might be suitable. The program director or education coordinator may have laboratory bench responsibilities as well. Since the classes are small, the type of teaching is a little different from the teaching done in an academic setting.

Research

Most positions in research will be found in industry, a university, or academic health center. Industrial research may involve new product development and performance of the necessary testing prior to product distribution. Research may involve the development or evaluation of a new laboratory method, a new clinical treatment method, a chemical, or pharmaceutical. Clinical laboratory scientists with a B.S. degree will generally be hired to perform the necessary testing for an already developed and funded research project. The principal investigator, who has developed the research project and protocol, is usually a Ph.D. or M.D.

Research positions require patience and honesty. Some weekend work or nights may be required, depending on the research project. Research scientists should be able to work independently and handle successes as well as failures in laboratory experiments.[5]

In addition to medical schools and universities, research institutions such as Wyeth, Smith-Kline, Wistar, and McNeil Laboratories hire clinical laboratory scientists. Good opportunities exist for clinical laboratory scientists in the area of biotechnology. Medical diagnostic industries are at the leading edge of technology-developing diagnostic and treatment products. Clinical laboratory scientists are the professionals of choice because they understand the end use of these new products and tests. Examples of products developed are closures for reagent packages, stoppers for blood collection tubes, and diagnostic tests for differentiating various types of tumors.[7]

Research provides in-depth expertise in a small area of scientific inquiry. Research may involve only a small amount of patient contact and interaction with others. The work may be repetitive and the position insecure if funded through grant or "soft" money.[5]

Management Positions, Administration and Supervision

Career opportunities exist for the highly motivated and experienced individual in directing laboratory staff and coordinating daily technical procedures. The minimum requirements are usually a B.S. degree and certification in the laboratory sciences. Positions are advertised under section chief, chief technologist, and laboratory operations manager.[7]

Do you aspire to a managerial post? Should you pursue time-consuming and expensive graduate study, such as an M.B.A program? Or should you undertake rigorous self-education? Training in managerial skills through continuing education programs is often available to would-be supervisors through the employer. Volunteer to complete certain tasks that your boss may have. Be willing to accept leadership positions, such as chairing a departmental committee on safety. Help out with inventory control, and update procedure manuals.[9]

Industrial and Commercial Positions

There are numerous opportunities in sales with companies manufacturing laboratory equipment or supplies. Clinical laboratory scientists are recruited since they are familiar with the equipment and laboratory testing. Since their customers are clinical laboratory scientists, they can converse with them easily. Characteristics that are needed to make a good salesperson include self-motivation, personality, creativity, energy, self-discipline, and integrity.[6]

Most companies would like their salespeople to have some background in sales or marketing as well as a laboratory background. If you wish to enter sales, consider taking some business and marketing courses. It is also helpful to have previous work experience in sales, such as in a department store, or fast-food chain.

Salaries are structured in a variety of ways. Some are straight commission (based on sales), some are straight salary, and others are combination of the two. Fringe benefits are usually good. In addition to health insurance, and life insurance, a company car might be provided. The company may support further education through tuition reimbursement. Company profit sharing may be another benefit.

It may sound very "romantic" to do a lot of traveling at someone else's expense and spend time away from home. However, it can be difficult on family life. There is a certain amount of stress and pressure related to meeting "sales quotas." Although a salesperson may plan his own hours, he must be versatile with time to consider his customers' availablilty.

In addition to sales positions related to laboratory equipment and reagents, there are other sales positions available not directly related to the clinical laboratory, such as in pharmaceuticals.

For those individuals who don't like sales but enjoy customer interaction and travel, a position as a technical field representative might be appealing. Technical field representatives serve as customer support staff. They may be responsible for customer training, continuing education, and solving instrumentation problems. These positions are usually salaried. The ability to make effective presentations to hospital staff, laboratory personnel, and physicians is essential.

A clinical laboratory scientist with an M.B.A can advance from product specialist to product manager or director of marketing and sales.

National corporations, such as General Electric, Johnson & Johnson, and Campbell's Soup perform both routine and research testing on their employees and sometimes clients, as well as testing for substance abuse. These employee health services are usually located on site.[7]

An interest in art, writing, or other creative field can be combined with the scientific skills of a clinical laboratory scientist. Clinical laboratory scientists with demonstrated writing abilities are needed to write technical reports and interpret complex bodies of data. Technical writers work for book publishers and editors of professional journals. The responsibilities of clinical technical writers include writing information bulletins, brochures, and newsletters. An artistically talented clinical laboratory scientist may be interested in working as an illustrator for publishers of medical journals, textbooks, and other publications.[7]

Organizations and firms that produce laboratory products also require the skills of a public relations director. Creativity, good communication skills, and diverse knowledge of the health care professions are essential.

An industrial hygienist or occupational health and safety (OS&H) officer generally requires additional education. There is a more gradual way to enter the field by becoming an occupational safety and health technologist through work experience.

An occupational safety and health technologist does safety inspections, monitors the levels of various chemicals or physical hazards, trains employees in health and safety issues, and records occupational accidents and illnesses.[2]

It is not necessary to work full-time in a safety and health position to become an OS&H technologist. People spending at least 35% of their work time on occupational safety or health activities are eligibile for certification after five years of experience. Successful completion of certain college science courses allows a person to be eligible after four years of work experience.[2]

Clinical laboratory scientists have a big advantage in such positions, particularly since quality control is a very important aspect of industrial hygiene testing. Laboratory professionals are aware that the quality of the samples taken determines the quality of the results.[2]

Major chemical and natural resource companies have challenging opportunities in chemistry and microbiology. High salaries are available in the Gulf Coast region and overseas.

Military and Government Positions

There are many advantages to positions with government agencies: career advancement, solid job security, and the possibility of travel or relocation. Federal, state, and local governments, as well as government-related agencies, are becoming increasingly involved in areas of health care.[7]

There are three branches of the armed forces that include laboratory personnel: the Air Force, Army, and Navy.[5]

Biomedical laboratory officers may manage, supervise, and perform laboratory tests in a variety of settings including hospitals and research laboratories. They also teach in medical/clinical laboratory sciences. Those wishing to be an officer should have a baccalaureate degree and have passed a national certification examination.[5]

All officers have the opportunity to apply for career-broadening assignments. Laboratory officers have held positions in systems, medical intelligence, recruiting, and as directors of ambulatory services.[5]

Continuing medical education (CME) classes and tuition assistance toward graduate education classes are available. Some of these classes are offered on base through local universities. There are opportunities for advancement in laboratory management, laboratory education, and specialty areas. Many of these positions are not dependent on someone leaving.[5]

A clinical laboratory scientist with a baccalaureate degree, certification, and no experience usually enters the armed services as a second lieutenant. Entry-level duties start at the basic supervisory/manager level, unlike civilian hospitals that would require previous work experience and advanced education for a supervisory position. Clinical laboratory scientists also provide professional consultation to the medical staff.[5]

Advantages of the military are job satisfaction, an esprit de corps often unknown in other areas of the military life, a chance to assume increased responsibilities, a wide variety of job and education opportunities, good pay and benefits, and a chance to see different parts of the country and the world. Military salaries are equal to those for comparable civilian positions, and military life offers many other inducements. Officers of the medical department of the armed forces carry equal rank and privileges with other military officers. There is the option to retire from the military after 20 years of service with excellent retirement benefits and begin a second career without loss of military retirement pay and benefits.[5, 6]

Government positions, other than the military, are available to clinical laboratory scientists. There are opportunities at the bench, in research, and in management at the Centers for Disease Control (CDC), in Atlanta. Positions also are available within the CDC and the Department of Health and Human Services for collecting and analyzing statistical data. These individuals are known as biometrists and have usually completed graduate coursework in statistics and research methodology.[5, 6]

Opportunities exist within the U.S. space program. A NASA mission specialist astronaut requires a master's degree and two years of experience in a specific specialty area, such as microbiology or hematology.[5]

CAREERS REQUIRING FURTHER EDUCATION

Graduate Education, Medical School

Many clinical laboratory scientists remain at the baccalaureate level. Ten percent or fewer graduates of CLS programs enter graduate schools. The highest percentages of CLS graduates have entered programs in medical school, business administration, education, clinical laboratory science, health-care administration, and microbiology. Other graduate programs that clinical laboratory science graduates have entered are law, engineering, medical physics, radiation science, public health, and optometry.[1, 5]

In many cases, the roles clinical laboratory scientists may assume in the future will require preparation at the graduate level. For example, positions as laboratory managers may require a master's degree in administration. An undergraduate degree in Clinical Laboratory Science is an excellent stepping stone into graduate school or medical school. Depending on the graduate program, there may be additional course requirements. Particularly in business, it may be necessary to complete undergraduate courses in business, finance, and accounting prior to starting graduate school. If you are pursuing an advanced degree in the sciences or intend to enter medical school, additional mathematics and physics courses may be required. A survey of graduate school program representatives by Beck and Chamness indicates that CLS graduates would be perceived as equivalent to other majors if they had the prerequisite courses and met other admissions criteria.[1]

Most graduate schools would prefer to have students who have an undergraduate grade point average of 3.00 (B). Some programs require students to take a national test such as the Graduate Record Examination (GRE).

Medical, Dental, and Veterinary School

Minimally, medical schools require students to have completed three years of college. Most prefer to accept students who have completed a B.S. degree, preferably in the sciences. Medical schools require a minimum number of credit hours in the sciences. Acceptance into medical school is based on grade point average, completion of the application form, acceptable Medical College Admission Test (MCAT) scores, and a formal interview. Work experience and extracurricular activities also are considered. Dental school and veterinary school admissions criteria are usually similar.

A number of clinical laboratory scientists often enter the specialty of pathology after completion of medical school. Their background in a clinical laboratory scientist is valuable.

In summary, clinical laboratory professionals possess concrete scientific and technical skills, organizational abilities, and the follow-through necessary to be successful in new adventures. There is no excuse for boredom or feeling "locked into" a particular position. There are always challenging opportunities available, if you are willing to adapt and take risks.

REFERENCES

1. Beck SF, Chamness MS: Graduate education for clinical laboratory scientists. *Clin Lab Sci* 1990; 3:49.
2. Campbell S: On-job training eases transition to OH&S technologist position. *Advance* 1990; 2:13.
3. Fox S: Career in academia: Is it right for you? *Advance* 1990; 2:1.
4. Gregg DE, Soldan AF, Hutchinson CS: Careers for clinical laboratory scientists: Verification of numerous and diverse opportunities. *Clin Lab Sci* 1990; 3:32.
5. Karni K, Oliver SJ: *Opportunities in Medical Technology Careers.* Lincolnwood, Ill, VGM Career Horizons, 1990.
6. Lindberg DS, Stevenson Britt M, Fisher FW: *Williams' Introduction to the Profession of Medical Technology,* ed 4. Philadelphia, Lea & Febiger, 1984.
7. Neibauer BC: Focus on nontraditional career opportunities: An opportunistic view. *Lab Med* 1987; 18:786.
8. Solberg P, Karni KR: Opportunity and versatility with a degree in clinical laboratory science. *Clin Lab Sci* 1989; 2:49.
9. Umiker WO: Now is career decision time for MTs. *Med Lab Observ* 1986; 18:47.

Chapter 12

Professional Ethics

PROFESSIONALISM AND OBLIGATION

Profession and *professionalism* have been defined in such terms as specialized knowledge, prolonged training, and services to the community. Profession, in the medical context, has been characterized by autonomy through a process of political negotiation and persuasion. Profession, however, transcends the technical activity itself to serve a moral end. It implies dedication to an ideal and, in medicine, professes the ability to perform "a good act of healing in the face of the fact of illness."[10]

CODE OF ETHICS OF THE AMERICAN SOCIETY FOR MEDICAL TECHNOLOGY[1]

Preamble

The Code of Ethics of the American Society for Medical Technology (ASMT) sets forth the principles and standards by which clinical laboratory professionals practice their profession.

The professional conduct of clinical laboratory professionals is based on the following duties and principles:

I. Duty to the Patient

Clinical laboratory professionals are accountable for the quality and integrity of the laboratory services they provide. This obligation includes continuing competence in both judgment and performance as individual practitioners, as well as in striving to safeguard the patient from incompetent or illegal practice by others.

Clinical laboratory professionals maintain high standards of practice and promote the acceptance of such standards at every opportunity. They exercise sound judgment in establishing, performing, and evaluating laboratory testing.

Clinical laboratory professionals perform their services with regard for the patient as an individual, respecting his or her right to confidentiality, the uniqueness of his or her needs, and his or her right to timely access to needed services. Clinical laboratory professionals provide accurate information to others about the services they provide.

II. Duty to Colleagues and the Professions

Clinical laboratory professionals accept responsibility to individually contribute to the advancement of the profession through a variety of activities. These activities include contributions to the body of knowledge of the profession; establishing and implementing high standards of practice and education; seeking fair socioeconomic working conditions for themselves and other members of the profession, and holding their colleagues and the profession in high regard and esteem.

Clinical laboratory professionals actively strive to establish cooperative and insightful relationships with other health professionals, keeping in mind their primary objective to ensure a high standard of care for the patients they serve.

III. Duty to Society

Clinical laboratory professionals share with other citizens the duties of responsible citizenship. As practitioners of an autonomous profession, they have the responsibility to contribute from their sphere of professional competence to the general well-being of the community, and specifically to the resolution of social issues affecting their practice and collective good.

Clinical laboratory professionals comply with relevant laws and regulations pertaining to the practice of clinical laboratory science and actively seek, within the dictates of their consciences, to change those that do not meet the high standards of care and practice to which the profession is committed.

> As a clinical laboratory professional, I acknowledge my professional responsibility to:
> Maintain and promote standards of excellence in performing and advancing the art and science of my profession;
> Safeguard the dignity and privacy of patients;
> Hold my colleagues and my profession in high esteem and regard;
> Contribute to the general well-being of the community; and
> Actively demonstrate my commitment to these responsibilities throughout my professional life.[1]

DEFINITIONS AND THEORIES OF MORALITY

"Moral" is derived from the Latin *mos* meaning custom or habit. "Ethics" is derived from the Greek *ethos* meaning custom, habitual use, conduct, and character. Today these two words are used somewhat interchangeably to refer to behavior in moral acts and include the notion of approval or disapproval of a given conduct.[12]

The question of right or wrong has bothered humankind from earliest times. Those in medical professions are confronted daily with decisions that must be approached using ethical concepts.

The social and ethical problems accompanying direct patient care are more remote to laboratory personnel than they are to physicians and nurses. Laboratory professionals may believe that their work is free of ethical problems associated with patient care, but ethical dilemmas frequently arise in the laboratory that either directly or indirectly affect patient care.[2]

Traditionally, laboratory professionals have resolved ethical dilemmas using conscience or instinct as guides. An ethical decision-making model provides a structured and more effective way to analyze and resolve an ethical dilemma.[2]

To some extent, even ethical analysis will not eliminate the frustrations associated with dilemmas in the laboratory. A true dilemma involves competing values, rights, interests, or duties that cannot be satisfied by a single action. Dilemmas do not have perfect solutions. The process of ethical analysis is one of ranking or establishing priorities for options.[2]

A perplexing dilemma may be approached by means of one of the broad ethical theories. The consequentialist, or utilitarian, theory focuses on consequences or outcomes, often emphasizing principles or actions that maximize beneficial outcomes. They solve problems in a way that will yield the greatest good for the greatest number of people.

The concept of patients' rights stems from Kantian philosophy; this approach, by emphasizing individual rights, strives to protect the minority or individual from the majority.[3]

Another approach, the formalist theory, pays less attention to the outcomes of an ethical dilemma than the form of behavior taken to deal with an ethical dilemma.

Choosing priorities is also heavily influenced by what one considers the fundamental principles guiding ethical behavior. A few commonly cited principles are:

- *Beneficence*—Principle of bringing about good or benefit whenever possible.[2]
- *Nonmalfeasance*—Principle of refraining from harming others.[2]
- *Autonomy*—Principle of encouraging freedom of choice and discouraging coercion or manipulation of others in an ethically troublesome situation.[2]

Other principles include treating people respectfully and equally, distributing benefits and burdens fairly, and treating similar cases in a similar manner.[2]

Circumstances may arise in which dissonance occurs between two principles, both of which are accepted as "good." This is called contingent incompatibility. For instance, a patient may wish to die with dignity, and it may be wrong to force him to live; this conflicts with the duty to prolong life.[3]

Kohlberg, a well-respected ethicist, describes three levels of moral development and decision making. The adult whose value system is in the punishment-obedience orientation of the infant is said to be in the preconventional level. In this stage decisions are made based on set rules. Punishment is expected if these rules are not followed. At the conventional level individuals are more concerned with social approval. The individual tends to conform to societal expectations such as upholding laws, contributing to society, and fulfilled duties. The highest level of moral development involves more internal and autonomous modes of thinking. As the individual progresses to this level, decisions will be determined by self-chosen ethical principles.[12]

Every decision made and every course of action taken is based on consciously and unconsciously held beliefs, attitudes, and values. Defining or clarifying these values is the first step in ethical decision making. The three processes in determining values are (1) choosing values; (2) prizing values; and (3) acting on these values. These values should be chosen from alternatives by thoughtful process and not by indoctrination.

A DECISION-MAKING MODEL

   I. Determine values.
  II. State the problem.
 III. List the alternatives.
  IV. Make a choice.
   V. Frame an ethical statement.
  VI. List the consequences.
       A. Immediate
       B. Long-term
 VII. Compare the consequences with a list of personal values.

This tool provides a workable framework on which to begin.

AN APPLICATION OF THE DECISION-MAKING MODEL

Personnel on the evening shift in the hematology laboratory of the hospital had gone home one Saturday evening. Only the senior technologist remained to finish some of his paperwork. Two technologists on the midnight shift arrived. One of the two was a young woman with little experience in the laboratory. When she approached the evening technologist to ask a question, he noticed that she seemed inebriated. Her gait was unsteady, and her speech was slurred. The evening laboratorian had heard rumors that this woman had come to work in an impaired condition before and suspected that she had been imbibing alcoholic beverages. He also knew that this collegue was not well liked by the rest of the staff who claimed she was careless and unable to handle complex problems.[2]

USING THE DECISION-MAKING MODEL

   I. *Values.* Excellent patient care and accurate correct laboratory testing are values that the evening shift supervisor holds. He may place value on the friendship and maintaining a good working relationship with the midnight technologist. He may need to determine the value he places on his free time vs. his need for a good night's sleep.
  II. *Problem.* The problem was what action the evening technologist should take to prevent harm to the patients and the employee.
 III. *Alternatives.* The evening technologist could have stayed for a double shift, assigning the impaired technologist to do harmless tasks. He could have warned the other midnight technologist of her co-worker's condition. He could have notified the weekend supervisor at home and asked for a decision on what to do. He could have done nothing and hoped for the best.[2]
  IV. *Choice.* He decided to do nothing and hope for the best.
   V. *Ethical statement.* It was the technologist's decision to do nothing. He would only take action when he had more proof. Concern for the other, perhaps falsely accused employee, was priority.

VI. *Consequences.*
   A. Immediate
      1. Evening technologist could go home on time and not work a double shift.
      2. Another supervisor would not have to be disturbed.
      3. The inebriated employee could continue to work, with no reprimands.
   B. Long-term
      1. Laboratory errors that could affect patient care.
VII. *Consequences and Personal Values.* The evening technologist's personal values centered upon himself and the other employee. There would be few immediate consequences to himself and the other employee by taking no action. The personal value of concern for the patients was less important. To consider: In the case of behaviors that both society and professionals recognize as illness, it is not just to differentiate between individuals on the basis of their profession; that is, treating one while punishing another. The chemically dependent person is just that—a person—deserving and in need of a large measure of care and concern, nonmalfeasance, and compassion, whether patient or professional.[6]

CONFIDENTIALITY

A nurse, an aide, and a clinical laboratory scientist learn that the local mayor is in the hospital with syphilis. The mayor had been admitted for another reason, but syphilis was detected through tests. During a lunch-table conversation with a group of co-workers, one person pipes up "Did you know the mayor is being treated for syphilis here?"[5]

Put yourself at the lunch table. Would you stop the conversation right then? What if the person breaching the confidence is a good friend? Would you risk alienating these health professionals by being righteous and acting like "Miss Manners"? Would you silently tolerate the conversation? Would you speak a cautionary word to your friend in private later?

The frequency with which casual breaches of confidentiality occur in hospitals suggests that relatively few health care professionals regard confidentiality as a matter of serious moral concern. A number of factors contribute to this laxness. First, breaches are almost never motivated by malice. Second, the person who breached confidentiality will probably be unaware of any harm that may result or of his or her role in having caused it. Third, such harm likely will consist of feelings of shame and embarrassment and not physical injury.[5]

Under the "old" health care system, the individual physician or nurse was directly responsible for protecting confidential information pertaining to the patient. In the "new" health care system, in a large hospital 75 people may have a legitimate "need to know" that warrants their receiving access to the patient's record. It is suggested that a "need to know" criterion should be used, to fix the range of individuals who have access to patient records.[5]

Health care professionals should not be intimidated by persons who demand patient information but have no clear right to it—including priests and ministers, police, lawyers, anxious friends, relatives, and other health care professionals not directly involved in the patient's care.

"The patient has the right to every consideration of privacy concerning his own medical care program . . . Those not directly involved in his care must have the permission of the patient to be present."[5]

AIDS AND ETHICS

AIDS is a challenge to critically review the foundations of social, professional, and personal ethics.

AIDS is a terminal illness. The underlying immune defect has never been shown to be stably reversible and almost all patients die within three years of their diagnosis. Although a fatal disease is usually known according to how it kills, AIDS was first known for whom it killed. Homosexual and bisexual men and intravenous drug users still comprise the largest segment of AIDS patients. Other victims include sexual partners of those with AIDS, prisoners, prostitutes, recipients of blood and blood products, people from Central Africa and Haiti, and children of all these persons. AIDS is no longer—if it ever was—exclusively a "homosexual" disease.[13]

There is an ethical danger that prejudice and panic may shape behavior toward AIDS victims. A primary ethical challenge is on the level of perception—to see the sufferer as greater than his disease and disability and certainly possessing a human dignity.

Ethical issues include basic attitudes toward AIDS and its sufferers and public perceptions about resource allocation for effective prevention of HIV infection. Ethical decisions may include the following:

1. Patients and care-givers must make decisions about initiating or continuing aggressive life-sustaining procedures.
2. Patients may become confused or mentally incompetent at the very time that ethically difficult choices must be made, such as use of experimental procedures.[9]
3. Patients may request that certain family members or friends not be told of the diagnosis of AIDS. The issue of confidentiality of medical records must be faced. There is one general guiding principle—the principle of "need to know." Access to confidential information should be limited to those who have a valid reason for having it.[4]

RISK AND OBLIGATION

For a variety of reasons, AIDS has been seen as a greater threat by health care professionals and by the lay public than is currently warranted. The public, fearful of risking AIDS by contributing blood, has curtailed blood donations. Some health professionals, fearful of contracting the disease, have hesitated to perform patient care, even care that carries only the most remote risk. At times, physicians have refused to take care of AIDS patients, surgeons have refused to operate on them, and pathologists have refused to perform autopsies on them. Fears associated with "catching AIDS" have varied from the slight, but realistic, fear of direct inoculation with blood to the mistaken notion that AIDS can result from merely touching patients.[10]

Risks to health care professionals are extremely small. Only two clear-cut cases of infection of a health care worker have been reported.[4]

In dealing with infectious disease, physicians have generally felt themselves disposed to assume the risks and to stay with their patients, and so, in more recent times, have other health professionals.[10]

AIDS, although often perceived differently, is an infectious disease, transmissible only under certain well-specified circumstances. It differs from other infectious disease in that (1) once transmission has occured, the certainty of clinical infection is questionable; (2) once the disease is actually diagnosed, it is almost always fatal; and (3) it carries with it a stigma.[10]

The backbone of medical professionalism—the willingness to exercise technical skills in pursuit of moral ends—implies three obligations: (1) the presence and maintenance of skills; (2) a conception of moral ends; and (3) a willingness to make moral choices and to engage in the activities appropriate to those choices.[10]

Individuals are not expected to pit themselves against certain death and they are not expected to assume risks forever: There are "reasonable" risks and there are acceptable ways of terminating professional obligations no longer suitable to a person or a condition.

It is concluded that health professionals have a historically grounded obligation to assume "reasonable" risks in dealing with AIDS patients, that "reasonable" is defined by the community, and that this modifiable obligation emerges from views of community, justice, and professionalism.[10]

LEGAL ISSUES AND AIDS

Laboratory and hospital managers must protect themselves against legal action for negligence, libel, slander, failure to obtain consent, breach of confidentiality, failure to warn, or failure to maintain a safe workplace.

The first problem is mistakenly supplying HIV-positive blood for transfusion. Most states consider blood products and blood units a service. In order to prove negligence, the plaintiff must demonstrate that there was a standard of care, that there was a breach of that standard of care, that an injury resulted, and, as a consequence, that damages should be paid.[7] Another possibility is a corporate liability case against a hospital or health care provider when a patient who has not been transfused is infected with HIV. Such situations might arise in patients undergoing dialysis for end-stage renal disease, since it is difficult to keep the filtration process infection-free.

If a patient is informed that he or she has AIDS, and the patient is able to prove that the test was unreliable or improperly performed, damages may be sought.[8]

Results reporting is another major AIDS testing issue. Hospital administration must devise a standard protocol, describing the steps the laboratory and the physician should take when a patient has a confirmed positive HIV result.

It is probably appropriate for those on the health care staff who might be affected to know that a patient is suspected of carrying the AIDS virus. However, it is recommended that all blood and body fluid specimens be treated carefully, with universal precautions. These precautions may include the use of gloves, goggles, and the proper disposal of the specimens upon completion of laboratory testing.[8]

SITUATIONS: ETHICS

As medical technology advances, more situations occur in which ethical decisions are required. Some of these are made by physicians, patients, or family members. Laboratory personnel, however, may have an impact on the final decision. Laboratory personnel also face dilemmas as they perform some laboratory testing.

These dilemmas may have more than one solution, some of which may be more acceptable than others. Consider the following:

Genetic Engineering.—It may soon be possible to correct chromosomal defects, such as sickle cell anemia and chronic granulocytic leukemia.

Research, Drug Testing.—If patients show improvement with a drug, should those patients on the placebo (medication with no therapeutic effect) also be given the drug? How much information should patients be given prior to entering these studies? What are your views on the use of animals in drug testing?

Euthanasia, Right to Die.—Do patients have the right to request that no unusual life support be used, such as respirators? Do patients have the right to request no liquids or nutrition?[11]

Child Abuse.—What are your responsibilities as a health care worker if you suspect child abuse on a patient brought in for laboratory tests?

Drug Use.—What do you do when a colleague steals drugs or uses drugs?

Informed Consent.—How much information should the patient be given about laboratory tests prior to blood drawing? How much information should be given prior to surgery or an experimental procedure?

Right to Strike.—Do health professionals have the right to strike even if this would affect patient care?

Workers at Risk.—If toxic substances are discovered at a work site, such as asbestos, what follow-up is necessary?

Organ Procurement, Transplantation.—With limited organs available for transplantation, what factors are used to determine who will receive these organs? What steps are necessary to assure the patient's and family's consent?

Blood Transfusions.—Do adults of certain religious sects, such as Jehovah's Witnesses, have the right to refuse blood transfusions? Do parents have the right to refuse blood transfusions for their children? In most cases, adults of sound mind do have the right to refuse blood transfusions. Children may be made wards of the court. At that time, they may receive blood transfusions without parental consent.[2]

Transfusion Reactions.—If you are investigating a transfusion reaction and determined that a certain employee, a friend, made the error, what would you do?

Results Reporting.—What can you do when a patient calls on the telephone requesting information on a laboratory test? How can you be sure that it is the patient?

Dialysis Patients.—With limited resources, how do you determine which patients will receive dialysis?

Workloads.—If excessive workloads at your institution result in more laboratory errors, what can be done?

Comatose Patients.—When drawing blood from comatose patients, do you assume the patient cannot hear you, and should you talk about him as if he were not there?

Physician Authorization.—If a certain procedure requires physician authorization, but the physician is unavailable, do you perform the procedure anyway?

REFERENCES

1. American Society for Medical Technology: *Code of Ethics.* Unpublished data, 1988.
2. Bingold JM, Malchiodi LR, Terry JS: Ethical dilemmas in the laboratory: The not-so-distant patient. *Clin Lab Sci* 1988; 1:230.
3. Chavigny KH: Ethical dilemmas and the practice of infection control. *Law Med Health Care* 1982; 10:168.
4. Desimone PA, et al: AIDS: Medical ethics (grand rounds). *Hosp Pract* 1986; 21:120.
5. Fleck LM: Confidentiality: Moral obligation or outmoded concept. *Health Prog* 1986; 67:17.
6. Fowler MD: Doctoring or nursing under the influence. *Heart Lung* 1986; 15:205.
7. Griffith J: AIDS: How to protect your lab on legal issues. *Med Lab Observ* 1989; 21:26.
8. Griffith J: How to defuse legal dynamite in the lab. *Med Lab Observ* 1986; 18:30.
9. Hansen RA: Ethics is the issue. *Am J Occup Ther* 1988; 42:279.
10. Loewy EH: Risk and obligation: Health professionals and the risk of AIDS. *Death Stud* 1988; 12:531.
11. McInerny WF: Understanding moral issues in health care: Seven essential ideas. *J Prof Nurs* 1987; 3:268.
12. McKinley MR: How do we decide: A model for making ethical decisions. *J Nephrol Nurs* 1984; 1:80.
13. Roy DJ, Tsoukas C: AIDS and clinical ethics: Honoring patients' dignity. *Dimens Health Serv* 1986; 63:32.

Chapter 13

Communication Skills and Public Relations

Why is communication important? "I didn't understand" is one of the most costly sentences in the clinical laboratory. It is costly in terms of time, money, and quality of care. Ninety percent of job failures are due to a breakdown in communication between people.[6]

Clinical laboratory scientists must possess not only technical expertise in their areas of specialization, but also a clear understanding of their professional, ethical, and moral responsibilities to clients and other health care providers. They should demonstrate a sensitive and caring interpersonal style. These affective dimensions of quality patient care emphasize how one talks to the patient and how one listens to the patient. In addition, health providers should exhibit empathy and warmth in a reciprocal and nonpatronizing relationship with a patient.

Communication is the process of passing information (written or oral) from one person to another. Interpersonal communication can be defined as a process of exchanging information and transmitting meaning between two individuals (dyadic communication) or by a small group of individuals. Organizational communication is a process by which managers develop a system to provide information and transmit meaning to large numbers of individuals within an organization and to relevant individuals and institutions outside the organization.[1, 9]

Communication skills are vital to developing and maintaining an effective team approach to health care. There are many individuals other than patients with whom a laboratory professional will interact. These include physicians, pathologists, laboratory supervisors, departmental supervisors, fellow laboratory workers, phlebotomists, nurses, ward clerks, other allied health professionals, and salespeople.

Communication influences morale and performance in health care facilities. Morale can be defined as attitudes and feelings of an individual toward his or her work. When morale is high, employees strive hard to accomplish the organization's goals. Conversely, low morale prevents or dissuades employees from pursuing the goals. The interpersonal communication skills of the clinical laboratory supervisor have a great effect on the morale of the laboratory staff. Communication also affects the emotional well-being of patients and their compliance with treatments and treatment plans. Caring feelings are essential in providing for the emotional needs of patients. Consider the positive effect that correct, calm explanations will have on a patient who is having blood drawn or undergoing other diagnostic procedures.[3]

Communication is a *dynamic, ongoing,* and *constantly changing interaction;* it is a two-way process. We must adjust and adapt to the constantly changing elements involved in human communication. Such adjustments and adaptations may range from modifying an explanation to a patient whose facial expression registers confusion, to meeting the needs and expectations of an audience during a public speech.[3]

Communication is *transactional* in nature. Participants in communication are constantly influencing one another. It is generally recognized that a speaker can influence a listener. When a speaker and listener are engaged in a communication transaction, the speaker can use verbal communication (words) and nonverbal communication (facial expressions, gestures, tone of voice) to influence the listener. A listener's means of influence is generally restricted to nonverbal communication.[3]

The *symbolic nature* of human communication also contributes to its complexity. Symbols stand for objects, actions, concepts, or feelings. Unfortunately, some of us assume that each word or symbol has a "right," "correct," or "complete" meaning. For example, seeing a familiar nonverbal symbol such as a white laboratory coat may mean that a health care facility adheres to a professional dress code. To many patients, especially the young, it means that they are about to get a shot or undergo a painful medical procedure.[3]

INTERPERSONAL COMMUNICATION: A MODEL

Figure 13–1 depicts the process of interpersonal communication as *dynamic, continuous,* and *complex*. The components include:[3, 8, 9]

1. *Source.* Interpersonal communication is generated by a source. The source, or speaker, generally wants to respond to his environment.
2. *Encoding.* The source (speaker) engages in what is known as encoding. Encoding takes place when the idea is transformed by the individual into a gesture, an action, a sound, written words, or a picture. The language chosen should be appropriate to the listener.
3. *Message.* The end result of the encoding process is the message. The primary function of a message is to express the purpose of the source. There is a message as perceived by the source, and there is the message as perceived by the receiver. This two-sided view is the root of many misunderstandings.
4. *Channel.* A channel is the medium by which a message is transmitted. There are many types of channels. They can be verbal (telephone calls or meetings) or nonverbal (letters, memoranda, and formal reports). Body language, visual communication, and silence also are channels. If something interferes with the channel, we call this interference *noise*. By "noise" we mean any kind of interference or distraction that comes between source and receiver. Noise, or channel interference, may result from distance, poor acoustics, faulty lighting, or a weak signal.[8, 9]
5. *Decoding.* The receiver (listener) must decode the message in such a fashion that he derives a meaning from it that is approximately the same as that transmitted by the source. Decoding is really encoding in reverse.[9] Differences in meaning occur because most sources

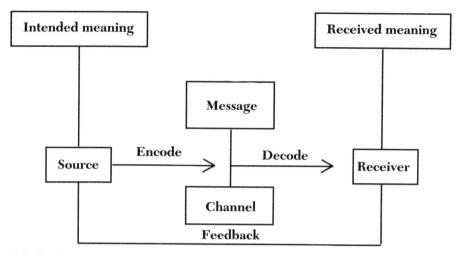

FIG 13-1.
Model of interpersonal communication.

and receivers have different backgrounds, experiences, values, needs, goals, expectations, attitudes, knowledge, and emotional makeups.

6. *Feedback.* Feedback refers to a response from the receiver (listener). Feedback is related to the circular or continuing nature of interpersonal communication. In the process of providing feedback, the receiver encodes and sends a message back through some channel to the original source. In this way, the source can tell whether the original message did get through to the receiver. If the message is unclear, the source might have to alter the encoding of the message or use a different channel.[9]

DEFINING EFFECTIVE COMMUNICATION

There are certain criteria the "good" communicator should keep in mind. These criteria are that the communication[8]:

1. Has a specific purpose
2. Is directed toward a specific audience
3. Is appropriate to the occasion or situation
4. Is timely
5. Is to the point
6. Is tactful
7. Is personal
8. Is clear and precise

Effective language may be described as that which is useful in a given situation.[8]

BARRIERS TO COMMUNICATION

The health professional may be from a socioeconomic group different from that of many of the people with whom he or she works. In the past, most health care personnel had middle-class socioeconomic backgrounds. In recent years, with increasing educational and employment opportunities, many members of lower socioeconomic groups and ethnic minority groups have moved into fields in which they were once only marginally represented.[3]

The language patterns one brings to the situation may affect relationships with patients and co-workers. In talking with individuals from the same cultural origin it might be desirable to use vernacular. However, if one continues to use only a minority pattern, some users of the majority pattern may react negatively. This can limit employability and the acceptance of the professional as competent. Ideally, we would wish for the members of our society to be more tolerant, but many people judge another's competency by the language patterns that the individual uses. If health professionals are aware of this, they may wish to develop more flexibility in their own language patterns.[3]

Other barriers to communication are (1) listener passivity; (2) wrong attitudes, such as annoyance and distress—shown by either the speaker or listener; (3) bad listening habits; and (4) the demands of time—speaking too fast, "rushing."[6]

Often, too, it is easy for the professional person to slip into the use of technical language or "jargon." Terms such as hemoglobin, intravenous, or CBC may be commonplace to you, but the use of these words may disconcert or alarm a patient. Technical terminology would be appropriate if the patient were another medical professional or perhaps a paraprofessional, but medical jargon to the average layperson is confusing. This also emphasizes that communication must adapt to the listener's background. A communication style in drawing blood from a child should be different from drawing blood from a nurse or fellow laboratorian.[3]

Characteristics of the listener, such as age, sex, education, physiological condition, social and economic background, and status need to be considered. One should not talk "baby-talk" to a 6- or 7-year-old. One should not shout and speak slowly to all older people, assuming them to be deaf.[3]

In a nutshell, ask "What type of person am I dealing with and how can I best get my message across to him or her?"[4]

PUBLIC RELATIONS

Laboratory personnel represent a health care agency as well as the laboratory. Communication weaknesses on their part have a much broader implication than a simple misunderstanding between a few people. The public has certain expectations relative to the images it feels its institutions should project. Even minor problems in communication, especially if such problems occur frequently, have a decided impact on public image.[8]

Although laboratory personnel are not as well recognized as many other providers of health care, good relations with the public and other health care professionals are in the laboratory's best interests. On all occasions a considerate and confident bearing must be maintained. Considerable time and energy are required to establish a laboratory organization that command the respect and confidence of the people who use the services.

Patients

Most patients have some degree of illness and discomfort about which their physician has expressed concern. Futhermore, these people are in strange surroundings and may suffer embarrassments and even indignities. They are assigned identification numbers and subjected to unfamiliar questions, examinations, treatments, and assorted other disturbing variations from their normal routine. They are apprehensive, insecure, frightened, and frequently feel threatened. It should be remembered that these patients are people, not numbers or tests. They are people in need of assistance.[7]

It is likely the patient will not greet the laboratory professional warmly for a blood draw. Health professionals should be friendly without being jolly or joking around. They should approach their work with confidence, and with a minimum of bustle. The patient is entitled to an explanation of what is going to be done but doesn't need detailed information on the reasons for the laboratory tests ordered by the physician.[3, 5]

Sometimes patients will attempt to get information through indirect questioning or coercion. They may demand that since they have paid for the tests, each has the right of ownership, and disclosure is mandatory. Such requests should be referred to the attending physician.[5]

Physician

The ultimate responsibility for the patient's well-being rests directly and singularly with the physician. The large assortment of clinical problems and the frequency with which they occur create a need for a range of laboratory services at various times and with various degrees of urgency. Only the physician can decide when tests are needed. Requests for services must not be met by complaints from "overworked" or "stressed out" laboratory personnel.

The sole laboratory response must be prompt acknowledgment of the desired services and their efficient and effective provision. Polite questions and suggestions are in order. Undesirable attitudes shown by laboratory workers to physicians, however, are unacceptable.[7]

Nursing

Nursing staff duties represent the hub of hospital patient care from which the needs for additional services are constantly generated. Because of the frequency with which these needs necessitate a laboratory response, a cooperative and coordinated relationship between the nursing and laboratory departments is imperative.[7]

CONFIDENTIALITY

Patients are interesting people. They sometimes have fascinating case histories and diagnoses. Two cardinal rules *must* be observed by all laboratory workers:

1. No interpretation of the results of the laboratory procedures is made to the physician or to the patient.

2. The information learned about patients is privileged, not public information. All such information should be held in strictest confidence. It should not be discussed in hospital elevators or in the hospital cafeteria![5]

IMPROVING COMMUNICATION

Improving communication depends on a number of skills. This includes listening effectively, demonstrating empathy, monitoring feedback, and recognizing nonverbal cues.

Listening.—Listening is the primary means by which we learn what we know. We spend about 70% of an average day engaging in verbal communication activities. Only about 30% of that time is spent talking, 16% reading, and 9% writing. The remaining 45% is spent listening. Research indicates that most individuals listen at about 25% efficiency. In other words, an average person experiences a 75% loss of information over a short period of time because of poor listening skills. Some factors that prevent good listening include lack of interest, contrary attitudes, information overloaded with details, interruptions, and mind-wandering. Remember that speech rate is slower than thought patterns.[8, 9]

There are four stages of listening: (1) hearing, (2) attention, (3) understanding, and (4) remembering. To improve listening stop talking, be tolerant, concentrate, control anger, and avoid drawing conclusions too early.[3]

Empathy.—Empathy is understanding the feelings as well as the content of another's message. It is seeing the world the way another individual sees it. Sympathy is not the same as empathy. Sympathy implies agreeing with what the other individual feels. Empathy requires no such judgment. Sympathy is often equated with pity, but empathy implies understanding.[8, 9]

Feedback.—There are many indicators or signals given to the source (speaker) by the receiver (listener) to indicate that the listener is understanding the message. Examples include facial expressions or posture, repetition, gestures, and tone of voice.

Other techniques also strengthen oral and written communications such as:

1. Verbal following (staying with the topic being discussed);
2. Using open-ended questions (those that start with "how," "why," "could," or "would") rather than closed-ended questions (those that can be easily answered by "yes" or "no");
3. Paraphrasing (stating in your own words the essence of what a speaker has just told you);
4. Reflecting feelings; and
5. Summarizing (pulling together, condensing, and clarifying the main points communicated to you).[6]

NONVERBAL COMMUNICATION

Imagine the feeling that you are captive, lying on your back in a 3-ft wide bed with no hope of escape. An individual enters and approaches with a tray full of tubes and bottles, and a snapping rubber tube casually flapping from a pocket. By his manner and facial expression this character may convince you that he is either bored with the whole procedure or is so happy to be there you wonder whether or not he is mentally competent.

"Put out your arm. Make a fist. Hold still. Don't move." Slap goes the alcohol sponge, pull and tie, and the tourniquet is in place. Out comes the 6-inch long needle with the quart-sized syringe or an array of at least a dozen tubes to collect your life-sustaining blood. This person has the responsibility of collecting specimens necessary for tests that were ordered by some obscure power possibly unknown to you. Remember that you were neither consulted nor advised and have no idea what is being done or why. You may wonder if this being that is working on you is human. Judging from the monotonous, staccato voice by which the instructions were delivered, you suspect it may well be a robot programmed to perform mechanically at the direction of some mysterious overseer. This exaggerated description of an everyday occurrence is included to introduce you to an integral part of the communication process—nonverbal cues.[8]

Nonverbal cues are "all stimuli other than spoken or written words."[8] These cues may depend on sight, hearing, and/or touch. If there is a discrepancy between the nonverbal and verbal message, the listener will generally respond to the nonverbal cue. Types of nonverbal messages include touch, body movement, eye behavior, eye contact, gestures, posture, jaw and mouth movements (frowns, smiles), and walk.

Voice intonation is also important. Voice qualities can indicate warmth, stress, or frustration. A lack of emotional sensitivity in voice tone can reduce trust between the speaker and the listener. Voice resonance, phonation, and articulation must also be considered.[8, 9]

Physical appearance and size is important. A large, imposing stature gives the appearance of authority. But a short or medium-height person can often obtain similar respect and authority through the use of posture and movement.[8]

The first impression that a health professional gives a patient can be very important. The clothing and accessories that you select make statements to others with whom you come in contact. Traditionally clothing has been a clue to the wearer's sex, age, status, and occupation. One intentional use of clothing as communication is in uniforms. The uniform may indicate status, the appropriate content for communication, and the type of interchange that may occur. Very short skirts and skin-tight pants are not appropriate in a hospital, clinic, or office. They can be distracting to fellow workers and patients. Dirty, down-at-the heels, and scuffed shoes, clumsy boots, and sandals are not very attractive nor supportive enough for people who must be on their feet several hours a day. The public may draw conclusions, rightly or wrongly, from the appearance of shoes and the care of hands and fingernails about the quality of care a facility or institution offers. In general, the health care professional should have hair that is clean, neat, and, if the hair is long, away from the face.

How people use the space around them communicates a great deal to others nonverbally. People have a need for personal space, which is an invisible "bubble" surrounding the person and moving with him from place to place. In various culture groups, the required personal space varies. In America, there are four boundaries of interaction: (1) the intimate (up to 2 ft);

(2) personal or casual (2 to 4 ft); (3) social or consultation (4 to 12 ft); and (4) public (12 ft or more).[8, 9]

COPING WITH SPECIAL PATIENTS

Laboratory personnel must insist on proper identification before collecting blood from a patient. In a hospital, the identification on the patient's armband should be checked against the information on the request/report form. In addition, the blood collector should ask all patients to identify themselves. If a patient is comatose or does not have proper identification, the phlebotomist should request the nurse to identify the patient. Verbal identifications should be noted on all request/report forms. In outpatient settings, the phlebotomist should ask the patients to identify themselves by spelling their names, perhaps. Asking "Are you Mrs. Jones?" is insufficient. Some patients might shake their head or say "yes" if they did not hear or understand you. Perhaps, there is more than one Mrs. Jones in the hospital![1]

Pediatric Patients

Five factors should affect communication with children. Consider the age of the child, if the child is separated from the parents, the child's understanding of health care experiences, and the interpersonal relationships already established in the health care environment. The most difficult children tend to be 7 months to 5 years old. They are often traumatized by the parents' absence and are subject to gross fantasies. They do not have strong reality concepts.

One should prepare the child for the experience. The child may perceive the blood-drawing experience as hostile, multilating, or a punishment for a wrong-doing. Take time to talk with the child and establish a trusting relationship. Be pleasant and relaxed. The environment may appear threatening to them. A child may see these strange instruments as "cold" and unfamiliar. Tell him why the procedure is to be performed, who will do it, and that it might be a little painful. Do not lie and tell the child it will not hurt. You'll never be trusted again! Do not talk to the parents in front of the child, ignoring the child. Give basic or simplified explanations.[3]

Elderly and Dying Patients

A terminally ill patient who has not yet been informed of his condition may ask if he is going to die. Try to be honest and direct, but do not go beyond the bounds of authority. This information is best given by the physician. You do want to be seen as a sensitive, caring individual. Forward the request to the appropriate individual.[3]

Recognize there are fears that dying patients experience: fear of abandonment, fear of pain, loss of independence, and fear of the unknown. Dying people progress through stages, as described by Kubler-Ross[3a]: denial, anger, bargaining, depression, and acceptance. Your communication style and content should be compatible with the stage of the patient.[3]

Do not talk about comatose patients as if they are not there or can't hear. Assume that any conversation you have with someone in the presence of the comatose patient may be heard and understood by the patient.

Uncooperative Patients

Uncooperative patients fall into two categories: those who are unable to cooperate and those who refuse to cooperate. Patients who literally cannot control themselves, due to a fear of needles or blood, are rare, but they do exist. Other patients may have uncontrollable movements, such as in Parkinson's disease. Yet another patient may have a paralyzed or contracted limb that he is unable to position for easy access to a vein. While these patients are unable to comply, this should not be misinterpreted as a refusal of the venipuncture. Both the patient and technologist generally want the blood to be drawn as quickly and painlessly as possible.

Patients who refuse to cooperate also fall into two categories: those who are incompetent and those who are competent. Children, the mentally handicapped, and the elderly with senile dementia lack decision-making capacities and belong in the first category. Adults may also have limited reasoning faculties when under the influence of intense emotions, drugs, or alcohol.

Any medical treatment that is performed without the patient's informed consent is a violation of the patient's rights. A surrogate, on behalf of a patient who lacks the capacity to make a decision, may make a decision with the best interest of the patient in mind. A parent may serve as a surrogate for a child or an adult child serve as a surrogate for a senile parent.[2]

TELEPHONE MANNERS[7, 8]

The health professional should be aware that people often form lasting judgments about people with whom they have communicated only by telephone.

1. Be courteous at all times.
2. Speak distinctly and in a pleasant tone of voice.
3. Be certain all questions and answers are clearly understood.
4. Do not answer questions about which you are uncertain; consult resources of the department (immediate supervisor, manuals).
5. Do not delay in answering the telephone.
6. Never leave the telephone unattended after a call is received; if placed on "hold," reassure the caller at frequent intervals that you are attempting to complete the connection.
7. When receiving calls, initiate the conversation with "Good morning (Good afternoon, Good evening), Department of Laboratories," or the appropriate greeting as determined by the supervisor.
8. When transferring calls, inform the person of the caller's name and department.
9. When making calls, preface remarks with your name and department.
10. Be certain all written or typed reports are neat, legible, accurate, dated, and initialed or signed.[7]

ORGANIZATIONAL COMMUNICATION SYSTEMS

The formal communication channels are established mainly by the organizational structure. Three formal types of communication are found in most organizations: (1) Downward communication; (2) Upward communication; and (3) Horizontal communication. The directions refer to the flow of information relative to the placement of the hierarchy of positions on an organization chart.[9] Downward communication flows from the supervisor to the subordinate. It generally takes the form of orders and is directive in nature. Often there is little request or interest for response from the lower levels.[1, 8, 9] Upward communication flows from the lower to higher authority. Upward communication is generally informative and reporting in nature. It may contain suggestions for decision making, or some form of protest over previous decisions.[1, 8, 9] Horizontal or lateral communication takes place between individuals who are approximately at the same level of authority in an organization. An example might be a communication about a patient blood specimen between the hematology and chemistry departments.[1, 9]

INFORMAL COMMUNICATION SYSTEMS

Informal communication systems exist but they cannot be identified by looking at authority relationships on a formal organizational chart. This informal communication system is frequently referred to as the "grapevine." It is an outgrowth of the social interaction of people and their desire to communicate. It allows subordinates to keep abreast of the latest information and thereby relieves their apprehension. It can be either beneficial or detrimental to an organization. It may provide insights into employee attitudes and helps spread useful information quickly.[1, 9]

As your career progresses, you will note that, for most positions, good communication skills are essential for good job performance.

REFERENCES

1. Becan-McBride K: *Textbook of Clinical Laboratory Supervision.* New York, Appleton-Century-Crofts, 1982.
2. Chilson PK: Restraining ethics considered. *Advance* 1990; 2:36.
3. Klinzing D, Klinzing D: *Communication for Allied Health Professionals.* Dubuque, Ia, Wm C. Brown, 1985.
3a. Kubler-Ross E: *On Death and Dying.* New York, Macmillan, 1969.
4. Kuhn PJ: Sharpening your communication skills. *MLO* 1987; 19:53.
5. Lindberg DS, Stevenson Britt M, Fisher FW: *Williams' Introduction to the Profession of Medical Technology,* ed 4. Philadelphia, Lea & Febiger, 1984.
6. Roseman E: Communication with your staff. *MLO* 1985; 17:89.
7. Shuffstall RM, Hemmaplardh B: *The Hospital Laboratory Modern Concepts of Management, Operations, and Finance.* St Louis, Mosby-Year Book, 1979.
8. Smith VM, Bass TA: *Communication for Health Professionals.* Philadelphia, JB Lippincott, 1979.
9. Snyder JR, Senhauser DA: *Administration and Supervision in Laboratory Medicine,* ed 2. Philadelphia, JB Lippincott, 1989.

Chapter 14

Future of Clinical Laboratory Science

PROGRESS IN HEALTH CARE

The most dramatic breakthrough in the health of Americans will probably be achieved, not in the field of acute care, but in prevention.[6]

The proportion of Americans 65 years and older increased from 4% in 1900 to 8.1% in 1950. This trend is expected to continue to 13% in 2000 and 21.8% in 2050. While the population as a whole will only double from 1950 to 2050, the number of Americans over 65 years will increase fivefold according to projections.[7]

The preventable diseases currently associated with an aging population include cancer, stroke, diabetes, hypertension, and cardiovascular disease. There will be many medical advances toward the prevention and treatment of these and other diseases. Research on AIDS and infertility will continue into the next century. More laboratory testing will allow monitoring of healthy individuals to promote health and prevent disease. An example is the increased cholesterol testing of Americans.[7]

In the 1990s, the public will be faced with decisions about health care costs. As new technologies and treatments occur, the costs of health care will rise. Health care providers will face a competitive market. The government will exert its influence through Medicare/Medicaid legislation and consideration of national health insurance programs.[6]

THE LABORATORY OF THE FUTURE

Today's clinical laboratory is being pressed on two fronts. A shortage of clinical laboratory scientists has made it difficult to get the work done. On the other hand, growing financial pressures have demanded that laboratories become more productive.[3]

The clinical laboratory of the 1990s will likely include the following:

Increasing computerization: test orders, specimen collection and processing, and test reporting will be computerized. There will be increased communication between laboratory departments by computer messages. Physicians will have greater access to computers to receive test reports quickly.[4]

Specimens will be identified by an electronic bar code reader. Positive patient identification will be made by matching the bar code on the patient's identification bracelet with the test slip and blood tube.[3-5]

Single instruments will offer a broad test menu for multiple laboratory disciplines. It is likely that instruments will perform tests according to technology, not specialty. These instruments are multidisciplinary test processors (MTPs). They can take a single specimen and perform a variety of tests linked only by some common handling and reading steps. An example of this type of instrument is the Kodak Ektachem. The integration of tests across traditional laboratory disciplines will ultimately eliminate the various laboratory sections as we know them.

Compact, easy-to-use instrumentation will allow testing to occur at the patient's bedside, the operating room, and other decentralized sites. An additional feature on future instruments will be the ability to appropriately order follow-up tests based on the result of the first test. Rather than wait for the clinician to review the first test and order follow-up tests, the instrument will handle this function. Laboratory personnel will have a new role monitoring this highly automated process.[3, 4]

Physicians will employ different testing strategies in the 1990s. There will be more emphasis on real-time testing or on-line monitoring of various chemicals. Testing will be selective, with few full profiles ordered. New profiles may be offered for cancer, opportunistic infections, and genetic abnormalities.[4]

The concept of autoflow—the abbreviated term for automated work flow—will be utilized. Bar-coded readers will be used to verify receipt and sort tubes according to workstation. Robots or conveyors may handle the transportation. Robots will play an increasingly important role in carrying out such mundane tasks as specimen sorting, receipt verification, and centrifugation. The ultimate step in robotics will be full automation of specimen flow from start to finish. It is anticipated that fewer (human) errors will be made when these tasks are performed by robots. In addition, they may be more cost-effective as wages for skilled help increase.[3-5]

In summary, we are heading toward a greater dependence on computers and advanced technology. They will reshape duties, instruments, and work flow in the laboratory of tomorrow.[4] There will be an increasing demand for laboratory personnel with knowledge of computers and advanced technologies.

SETTINGS FOR HEALTH CARE DELIVERY

Hospitals and private laboratories are currently the predominant employers of clinical laboratory scientists. The Commission on Allied Health Education has predicted that although more services are likely to be provided outside hospitals, the demand for hospitalization will rise in the near future.[6]

Hospital bed numbers will increase and the increase will be in health care conglomerates, which will include nursing homes. Many of the elderly in nursing homes will require ongoing laboratory testing. Hospitals and independent laboratories will compete for these markets by offering incentives such as courier services and phlebotomists for blood drawing on site.

Preoperative laboratory testing and follow-up testing after surgery will continue on an outpatient basis, as hospitals face cost-containment issues. In addition, personnel costs will be evaluated. In some cases, clinical laboratory scientists with B.S. degrees will be replaced by lower-level personnel.

Clinical laboratory scientists will be found in ambulatory care settings and nursing homes. Business and industry will continue to hire certified laboratory personnel as well.

IMPACT OF NEW TECHNOLOGIES

There has been a biotechnology explosion in the last ten years. Reports of discoveries and inventions have become commonplace, as have new applications for old techniques.[2]

Techniques previously seen only in highly sophisticated research laboratories can now be easily performed in the routine clinical laboratory, thanks to advances in instrumentation.[2]

During the 1990s, this trend will continue, and it will have a major impact on the clinical laboratory. Emerging technologies, such as DNA probes and nuclear magnetic resonance, will gain wider use. Other established technologies, such as immunoassays, will continue to evolve.[2]

The immunoassay diagnostics market is expected to top $1 billion by 1993, expanding more than sixfold from the $160 million recorded in 1987, according to Theta Corporation, a market research firm. The question now is how to move the technology into the clinical laboratory in an easily usable form.[2]

One way to simplify testing is to develop methods that use today's chemistry analyzers. Another trend is the use of random access analyzers able to perform a very broad menu of individual tests. Such instruments keep supply costs low and equipment expenses down, and make better use of personnel. These instruments will rely on specialized test packs containing all the materials, reagents, and substrate necessary for complex tests.[2]

By the 1990s, immunoassays will probably be performed no differently from any other routine test. Few tests will be performed in the traditional radioimmunoassay (RIA) laboratory. RIA tests will be converted to other immunoassay methods.[2]

Biosensor technology will continue to develop. A biosensor is a microelectronic device which uses a biological molecule, such as an antibody or enzyme, as a sensing element. It is well suited to decentralized testing sites and smaller laboratories. If implanted in a patient, it can provide real-time biochemical monitoring. An example would be monitoring of blood glucose levels for a diabetic patient.

Advantages of biosensors include simple and low-cost instrumentation, fast response times, and high specimen throughput. Disadvantages include frequent calibration requirements and the limited working lifetime of a biosensor.[2]

Recombinant DNA technology, commonly termed DNA probes, has extended from the research laboratory into the clinical laboratory. DNA probes are useful in identifying microorganisms and diagnosing diseases with a genetic basis, such as chronic granulocytic leukemia. Other uses of DNA technology include paternity testing and detection of oncogenes (genes producing tumors).

A highly specific probe can be made in the laboratory containing a specific sequence of nucleotides. Nucleotides are the "building blocks" of the DNA strand. One may extract DNA from a living cell and obtain a single stranded DNA molecule. The laboratory manufactured DNA molecule, the probe, can be added to the DNA extracted from the cell. If the sequences are complementary, the DNA strands will hybridize, forming a double-stranded DNA molecule. Thus, new DNA molecules are constructed in vitro.

Nuclear magnetic resonance (NMR) is a very powerful diagnostic tool. In the clinical laboratory, high-resolution NMR spectroscopy can be used to study different compounds. NMR spectrophotometry measures the spin of protons in a magnetic field after a radiofrequency current has been applied. NMR has many applications in toxicology and drug identification. Gas chromatography and mass spectrophotometry will continue to be used as well.

THE LABORATORY AND HEALTH CARE COSTS

It is likely that 15% of a family's income will be spent on health care in the 1990s. Laboratory tests account for nearly 10% of total national health care costs. Cost containment measures will include the following:

1. Laboratory utilization by physicians will be monitored by hospitals. Unneccessary patient testing is discussed with the medical staff.
2. Laboratory managers will be nonphysicians with a strong business and management background.
3. Laboratory instrumentation and methodologies will be critically evaluated for cost prior to implementation.
4. Physicians will access microcomputers for suggested initial laboratory testing and diagnosis. Suggested follow-up laboratory tests will be provided as well.[6]

PERSONNEL—SUPPLY AND DEMAND

According to Castleberry and Kuby,[1] historical and demographic data from the Bureau of Labor Statistics (BLS) demand projections suggest supply and demand will be further out of balance than it is today. In other words, the supply of qualified laboratory personnel will worsen in the decade ahead.[1]

In 1986, the BLS estimated that 239,000 clinical laboratory technologists and technicians (medical technologists and medical technicians) were employed.

Considering a 2% attrition rate and new graduates, it is projected that there will be a *supply* of 239,353 clinical laboratory technologists and technicians by the year 2000. That's just 353 more than in 1986.[1]

The most recent BLS projections cover the period 1986–2000. Low, moderate, and high employment projections were developed. The *demand* for technologists and technicians are 285,000, 296,000 and 307,000.[1] Figure 14–1 illustrates the supply-and-demand picture. The laboratory is already experiencing shortages in several personnel categories. Vacancy rates for qualified laboratory personnel vary from 5% to 13%, with staff cytotechnologists vacancy rates being the highest.[1]

The shortage of technologists is expected to be as acute as that now seen in nursing. Frequently mentioned reasons include other career opportunities for women traditionally entering the field and limited opportunity for advancement in status or salary. More specialized testing

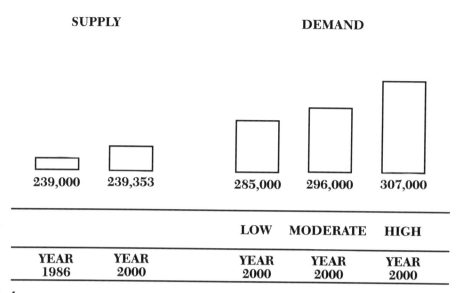

FIG 14-1.
Supply and demand for clinical laboratory technologists and technicians. Moderate level projection equates with 57,000 new positions by the year 2000, a 24% increase over 1986. (From Bureau of Labor Statistics.)

will require additional qualified personnel. These developing areas include cytogenetics, tissue typing, transplantation, gene-splicing techniques, and production of monoclonal antibodies.[6]

The providers of health care will be similar to those we currently have, although their numbers and specialties will change. Clinical laboratory scientists will be adding to the body of knowledge through research. Communication skills and management skills will become increasingly important. Technician-level skilled persons will operate many of the instruments or draw and process blood.[6]

LABORATORY EDUCATION

School closings continue at a significant rate. There were 639 accredited schools of medical technology in 1982. In 1988, there were 464 schools. Most schools that closed were hospital-based programs. The two main reasons for program closures are (1) high financial costs of the program and (2) lack of a qualified student applicant pool. Medical laboratory technician programs have not been affected as much, as many are academically based. The trend of educational programs shifting from hospitals to universities will continue. The trend for students to seek college and university enrollment and academic credit will continue into the 1990s.[1, 6]

We can expect to find greater use of educational technology. There has been increased use of educational television and teleconferencing. Microcomputers will continue to be used for tutorials and instruction.

Education of the future will need to stress more interdisciplinary involvement with other allied health professions, a greater emphasis on communication skills, increased knowledge of issues in health care, and the development of skills for initiating public policy research.[6]

REFERENCES

1. Castleberry BM, Kuby AM: Who will staff the laboratory of the 90s? *Med Lab Observ* 1989; 21:59.
2. De Cresce RP, Lifshitz MS: The impact of the new technology. *Med Lab Observ* 1989; 21:41.
3. Lifshitz MS, De Cresce RP: Automation: Trends in instrumentation, robotics, computers. *Med Lab Observ* 1989; 21:73.
4. Lifshitz MS, De Cresce RP: The clinical laboratory of the future. *Med Lab Observ* 1988; 20:30.
5. Pippenger CE, Megargle RG, Galen RS: The robots are coming. *Med Lab Observ* 1985; 17:30.
6. Ross DL: The laboratory environment: 1990–2000. *Am J Med Tech* 1982; 48:282.
7. Statland BD: How illness demographics will affect the lab. *Med Lab Observ* 1989; 21:79.

Appendix A

Glossary

Accreditation. A voluntary peer process whereby a private, nongovernmental agency or association grants public recognition to an institution or specialized program of study that meets or exceeds nationally established standards of acceptable educational quality.

Allied Health. A large cluster of health care-related personnel whose functions include assisting, facilitating, or complementing the work of physicians and other specialists in the health care system.

Career Ladder. Lateral or vertical steps that link jobs within the same job family and permit an employee, based on education and experience, to move to an advanced position or to a related occupation.

Certification. The process by which a nongovernmental agency or association grants recognition to an individual who (1) meets predetermined requirements, specified by that agency or association; and (2) voluntarily seeks such recognition.

Chief Executive Officer (CEO). The principal administrative official of an institution or agency.

Committee on Allied Health Education and Accreditation (CAHEA). An allied health education accrediting agency sponsored by the American Medical Association.

Council on Postsecondary Accreditation (COPA). A national, nonprofit private sector organization whose major purpose is to support, coordinate, and improve all nongovernmental accrediting activities conducted at postsecondary educational levels in the United States.

Hospital-Based Program. An educational program sponsored by an accredited hospital.

Joint Commission on Accreditation of Healthcare Organizations (JCAHO). An independent private agency that accredits hospitals and related, or similar, facilities.

Licensure. The process by which an agency of government (1) grants permission to persons meeting predetermined qualifications to engage in a given occupation and/or to use a particular title, or (2) grants permission to an institution to perform specified functions, otherwise prohibited.

Practicum. An extended period of full-time experience (weeks or months) in professional practice during which students reconstruct and apply the theory they have learned and during which clinical proficiency is developed. A practicum is a type of internship.

Program Director. The person in charge of developing and maintaining an educational program within an institution, hospital, or other sponsoring agency.

Registration. The process by which qualified individuals are listed on an official roster maintained by a governmental or nongovernmental agency.

Registry. A published list of those who are registered, or an agency which publishes a list of those registered.

Technician. One who specializes in the technical detail of a body of work; one who has acquired the ability to perform a complex task or set of tasks.

Technologist. One who specializes in the application of scientific knowledge to solve practical and/or theoretical problems; may supervise technician personnel.

University-Based Program. An educational program sponsored by an accredited university.

REFERENCE

1. Gupta GC: *Allied Health Education Directory, 1990*, ed 18. Chicago, American Medical Association, 1990.

Appendix B

Acronyms and Abbreviations

AAB	American Association of Bioanalysts
AABB	American Association of Blood Banks
AACC	American Association for Clinical Chemistry
AHA	American Hospital Association
AIDS	Acquired immunodeficiency syndrome
AMA	American Medical Association
AMT	American Medical Technologists
ASAHP	American Society of Allied Health Professions
ASC	American Society of Cytology
ASCP	American Society of Clinical Pathologists
ASM	American Society for Microbiology
ASMT	American Society for Medical Technology
BHPr	Bureau of Health Professions
CAP	College of American Pathologists
CDC	Centers for Disease Control
CLIA	Clinical Laboratory Improvement Act (1967, 1988)
CLS	Clinical laboratory scientist, clinical laboratory science
CLT	Clinical laboratory technician
CME (AMA)	Council on Medical Education of the American Medical Association
COPA	Council on Postsecondary Accreditation
DHHS	Department of Health and Human Services
DRGs	Diagnosis-related groups
ED	US Department of Education
HCFA	Health Care Financing Administration
HMO	Health maintenance organization
HSRA	Human Services and Resources Administration
JCAHO	Joint Commission on Accreditation of Healthcare Organizations
MT	Medical technology, medical technologist
MLT	Medical laboratory technician
NAACLS	National Accrediting Agency for Clinical Laboratory Sciences

NCA	National Certification Agency for Medical Laboratory Personnel
NSH	National Society for Histotechnology
PPS	Prospective payment system
SNM	Society of Nuclear Medicine
USPHS	United States Public Health Services
VA	Department of Veterans Affairs

REFERENCES

1. Gupta GC: *Allied Health Directory, 1990*, ed 18. Chicago, American Medical Association, 1990.
2. Karni K, Oliver JS: *Opportunities in Medical Technology Careers*. Lincolnwood, Ill, NTC Publishing Group, 1990.

Appendix C

Professional Organizations and Addresses

American Association of Bioanalysts (AAB) 314-241-1445
818 Olive St., Suite 918
St Louis, MO 63101

American Association of Blood Banks (AABB) 703-528-8200
1117 North 19th St., Suite 600
Arlington, VA 22209

American Association for Clinical Chemistry (AACC) 202-875-0717
2029 K Street NW, 7th Floor
Washington, DC 20006

American Association of Immunologists (AAI) 301-530-7178
9650 Rockville Pike
Bethesda, MD 20814

American College of Radiology 703-648-8900
1891 Preston White Dr.
Reston, VA 22091

American Hospital Association (AHA) 312-280-6447
840 N. Lake Shore Dr.
Chicago, IL 60611

American Medical Association (AMA)
515 North State St.
Chicago, IL 60610

American Medical Technologists (AMT) (see Certifying Agencies)

American Society of Allied Health Professions (ASAHP) 202-857-1150
1101 Connecticut Ave. NW, Suite 700
Washington, DC 20036

American Society of Hematology (ASH) 609-845-0003
6900 Grove Rd.
Thoroughfare, NJ 08086

American Society for Medical Technology (ASMT) 202-785-3311
2021 L St. NW, Suite 400
Washington, DC 20036

American Society for Microbiology (ASM) 202-833-9680
1913 I St. NW
Washington, DC 20006

American Society of Clinical Pathologists (ASCP) 312-738-1336
2100 W. Harrison St.
Chicago, IL 60612

American Society of Cytology 215-922-3880
1015 Chestnut St., Suite 1518
Philadelphia, PA 19107

American Society of Radiologic Technologists 505-298-4500
15000 Central Ave. SE
Albuquerque, NM 87123

Association of Cytogenetic Technologists, Inc. (ACT) 602-945-4363
for information on membership, contact:
  Helen Bixenman
  Membership Secretary
  Genetrix
  6401 E. Thomas Road
  Scottsdale, AZ 85251
for additional information on ACT, contact:
  Mike Zvarich, Public Relations Officer
  Genetics Associates of North Carolina
  120 Conner Drive, Suite 201
  Chapel Hill, NC 27514 919-942-0021

Canadian Society of Laboratory Technologists (CSLT)   416-538-8642
PO Box 2830 Station A
Hamilton, Ont. L8N 3N5

Clinical Laboratory Management Asso (CLMA)   215-647-8970
193–195 West Lancaster Ave.
Paoli, PA 19301

International Association of Medical Laboratory Technologists (IAMLT)
1 Drayton Gardens
Winchmore Hill, London N21 2NT, UK

International Society for Clinical Laboratory Technology (ISCLT)   314-241-1445
818 Olive St., Suite 918
St. Louis, MO 63101

National Society for Histotechnology (NSH)   301-577-4907
5900 Princess Garden Pkwy, Suite 805
Lanham, MD 20706

Society of Nuclear Medicine   212-889-0717
136 Madison Ave.
New York, NY 10016

ACCREDITING ORGANIZATIONS FOR LABORATORY EDUCATION PROGRAMS

Accrediting Bureau of Health Education Schools (ABHES)   219-293-0124
Oak Manor Office
29089 US 20 West
Elkhart, IN 46514

Committee on Allied Health Education and Accreditation (CAHEA)   312-464-4624
515 N. State St.
Chicago, IL 60606

Council on Postsecondary Accreditation (COPA)   202-452-1433
One Dupont Circle NW, Suite 305
Washington, DC 20036

National Accrediting Agency for Clinical Laboratory Sciences (NAACLS)   312-714-8880
Citicorp Plaza, Suite 670
8410 W. Bryn Mawr
Chicago, IL 60631

CERTIFYING AGENCIES FOR GRADUATES OF LABORATORY EDUCATION PROGRAMS

American Medical Technologists (AMT) 312-823-5169
AMT Building
710 Higgins Rd.
Park Ridge, IL 60068

American Board of Clinical Chemistry

Board of Registry of American Society of Clinical Pathologists (ASCP) 312-738-1336
PO Box 12270
Chicago, IL 60612

The Credentialing Commission International Society for Clinical 314-241-1445
Laboratory Technology
818 Olive St., Suite 918
St. Louis, MO 63101

National Certification Agency for Medical Laboratory Personnel (NCA) 202-857-1023
2021 L St. NW, Suite 400
Washington, DC 20036

National Registry in Clinical Chemistry

National Registry of Microbiologists 202-737-3600
% American Society for Microbiology
1325 Massachusetts Ave. NW
Washington, DC 20005

Nuclear Medicine Technology Certification Board 404-493-4504
Box 806
Tucker, GA 30084

Index

A

Abbreviations, 139–140
Accreditation
  agencies for, 55–57
  hospital classification based on, 10
  organizations for, 143
Accrediting Bureau of Health Education Schools (ABHES), 57
Acronyms, 139–140
Administration, employment opportunities in, 108
Administrative Technologists in clinical laboratory, 17
Advancement, 39–42
Agencies
  accreditation, 55–57
  certification, 51–54 (see also Certification, agencies for)
AIDS
  ethics and, 117–118
  legal issues and, 118
  risk and obligation related to, 117–118
American Association for Clinical Chemistry (AACC), 59
American Association of Bioanalysts, 59
American Association of Blood Banks (AABB), 59
American Association of Immunologists (AAI), 60
American Board of Clinical Chemistry (ABCC), 55
American Medical Technologists (AMT), 53
American Society for Medical Technology (ASMT), 57–58
  code of ethics of, 112–113
American Society for Microbiology (ASM), 61
American Society of Allied Health Professions (ASAHP), 60
American Society of Clinical Pathologists (ASCP), 58–59
American Society of Hematology, 60

Automatic analyzers, continuous-flow, introduction of, 7
Autonomy, ethical behavior and, 114

B

Beds, number of, hospital classification based on, 10
Beneficence, ethical behavior and, 114
Blood bank(s), 67–68
  laboratory technical positions in, 104
  technology for, specialists in, in clinical laboratory, 37
Blood banking, history of, 6
Blood cell count, complete, 65–66
Board of Registry of American Society of Clinical Pathologists (ASCP), 51–52
Bureaucratic principles of hospitals, 11–12

C

CAHEA program, 79–80
Career ladder, 39–42
Career opportunities, 83–111
  in administration, 108
  in commercial positions, 108–109
  in cytogenetics, 86–91 (see also Cytogenetics)
  in cytotechnology, 91–94 (see also Cytotechnology)
  in education, 106–107
  in government positions, 109–110
  in histology, 94–97 (see also Histology)
  in industrial position, 108–109
  in laboratory technical positions, 101–106 (see also Laboratory technical positions)
  in management, 108
  in military positions, 109–110
  in nuclear medicine, 83–86 (see also Nuclear medicine)

145

Career opportunities (cont.)
  outside health care, 106–110
  requiring further education, 110–111
  in research, 107
  in supervision, 108
Certification, 44
  agencies for, 51–54
    American Board of Clinical Chemistry as, 55
    American Medical Technologists (AMT) as, 53
    Board of Registry of American Society of Clinical Pathologists (ASCP) as, 51–52
    Department of Health and Human Services (DHHS) as, 54
    for graduates of laboratory education programs, 144
    independent, 54–55
    International Society for Clinical Laboratory Technology (ISCLT) as, 54
    National Certification Agency for Medical Laboratory Personnel (NCA) as, 53
    National Registry in Clinical Chemistry as, 54–55
Chemistry laboratory, 68–71
Chromatography, discovery of, 7
Clerks in clinical laboratory, 19
Clinical Diagnosis and Management by Laboratory Methods, 6
Clinical laboratory
  departments of, 64–72
    blood bank as, 67–68
    chemistry, 68–71
    coagulation, 64–65
    hematology, 65–66
    immunology/serology, 69–71
    microbiology, 71–72
    urinalysis, 66–67
  federal regulation of, 49–50
  of future, 131–132
  future health care costs and, 134
  licensure of, 48–49
  nontechnical health care-related professions in, 105–106
  organization of
    in hospitals, 15–19
    past, present, and future, 19–20
  physical facilities for, 19
  staff of, 32–37 (see also Staff, laboratory)
Clinical Laboratory Management Association (CLMA), 61
Clinical laboratory science (CLS)
  definition of, 1–4
  future of, 131–135

  history of, 5–8 (see also History of clinical laboratory science)
  medical technology versus, 2–3
  nature of work in, 1–2
  profession of, 4
  scope of practice of, 3
Clinical laboratory scientist (CLS), 76–81
  certification of, 80
  in clinical laboratory
    roles/functions of, 34
    training of, 19
  education for, 35, 76–79
    academic coursework in, 76
    programs for, 76–79
Clinical laboratory scientist generalist examination, requirements for, 80–81
Clinical laboratory technician (CLT), 73–75
  certification of, 74
  in clinical laboratory
    roles/functions of, 35
    training of, 19
  education for, 35, 73–74
  requirements for, 74–75
Clinical sciences specialists on laboratory staff, 33–34
Coagulation laboratory, 64–65
Code of Ethics of American Society for Medical Technology, 112–113
College of American Pathologists (CAP), 62
Commercial laboratory, laboratory technical positions in, 103–104
Commercial positions, employment opportunities in, 108–109
Committee on Allied Health Education and Accreditation (CAHEA), 56
Communication
  barriers to, 124
  confidentiality and, 125–126
  in coping with special patients, 128–129
  effective, defining, 123
  improving, 126
  interpersonal, 122–123
  nonverbal, 127–128
  public relations and, 124–125
  systems for
    informal, 130
    organizational, 130
  telephone manners and, 129
Communication skills, 121–130
Compatibility testing in blood bank, 67
Complete blood cell count (CBC), 65–66
Computers
  employment opportunities in, 106

introduction of, 8
Confidentiality, 116–117, 125–126
Cytogenetics, 86–91
   certification in, 87–88
   development of, 7
   educational preparation for, 87
   job opportunities in, 87
   occupational description of, 86–87
   professional organizations for, 88–91
Cytometry, flow, 104
Cytotechnologist in clinical laboratory, 17
Cytotechnology, 91–94
   certification in, 93
   educational preparation for, 93
   job opportunities in, 92
   occupational description of, 92
   professional organizations for, 93–94

D

Decision-making model, professional ethics in, 115–116
Dental school, careers requiring, 111
Department of Health and Human Services (DHHS), 54
Diagnosis-related group (DRG) coordinators, employment opportunities as, 106
Dying patients, communication with, 128

E

Education
   employment opportunities in, 106–107
   further, careers requiring, 110–111
   laboratory, future, 135
Elderly patients, communication with, 128
Emotional requirements, 22–23
Empathy in improving communication, 126
Employee turnover as challenge of 1990s, 26–27
Employment opportunities, 83–111 (*see also* Career opportunities)
Enrollment, student, declining, as challenge of 1990s, 25–26
Environmental laboratory, laboratory technical positions in, 104
Ethics, professional, 112–120 (*see also* Professional ethics)

F

Federal regulation of laboratories, 49–50
   increasing, as challenge of 1990s, 29–30
Feedback in improving communication, 126

Fibrinolysis, 64
Flow cytometry, 104
Forensic laboratory, laboratory technical positions in, 104
Future
   education in, 135
   health care in, 131
      costs of, 134
      delivery settings for, 132–133
   laboratory of, 131–132
   new technologies of, impact of, 133–134
   personnel in, supply and demand for, 134–135

G

Government positions, employment opportunities in, 109–110
Graduate education, careers requiring, 110–111

H

Health care
   costs of, future, laboratory and, 134
   delivery of, settings for, future, 132–133
   progress in, 131
   system of, U.S., 9
Health care personnel placement specialists, employment opportunities as, 106
Health care team approach in hospitals, 15
Hematology laboratory, 65–66
Hemostasis, 64–65
Histology, 94–97
   certification in, 96–97
   educational preparation for, 95–96
   job opportunities in, 95
   occupational description of, 94–95
   professional organizations for, 97
History of clinical laboratory science, 5–8
   post–World War II, 6–8
   prior to 1900, 5–6
   1900 to World War II, 6
Histotechnologist in clinical laboratory, 17
Hospital(s)
   bureaucratic principles of, 11–12
   classification of, 10–11
   clinical laboratory organization in, 15–19
   health care team approach in, 15
   informal organization of, 14–15
   laboratory organization and, 9–20
   laboratory technical positions in, 102–103
   organization chart for, 12–14
   professional service departments in, 15
Hybridoma technology, introduction of, 8

I

Immunoelectrophoresis, development of, 7
Immunohematology laboratory, 67–68
Immunology/serology laboratory, 69–71
Industrial laboratory, laboratory technical positions in, 103–104
Industrial positions, employment opportunities in, 108–109
Infection control, employment opportunities in, 105
Informal communication systems, 130
Informal organization of hospitals, 14–15
Information, sources of, 62
Integration, vertical, hospital classification based on, 11
Intellectual requirements, 22
International Association of Medical Laboratory Technologists (IAMLT), 62
International Society for Clinical Laboratory Technology (ISCLT), 54
Interpersonal communication, 122–123
In vitro fertilization laboratory, laboratory technical positions in, 104

J

Job satisfaction as challenge of 1990s, 26–27

L

Laboratory (*see* Clinical laboratory)
Laboratory Coordinators in clinical laboratory, 17
Laboratory Managers in clinical laboratory, 17
Laboratory science program, applying to, 81–82
Laboratory Scientists/Specialists in clinical laboratory, 17
Laboratory technical position(s), 101–106
 in commercial laboratories, 103–104
 in hospitals, 102–103
 in industrial laboratories, 103–104
 in other sites, 104–105
 in private laboratory, 103
 in public health laboratories, 104
 in veterans administration and other federal hospitals, 104
Legal issues, AIDS and, 118
Licensure, 44, 45
 laboratory, 48–49
 personnel, 45–48
  cons of, 46
  pros of, 45–46
  update on, 47–48
Listening for improving communication, 126

M

Management positions, employment opportunities in, 108
Manners, telephone, 129
Manpower shortage as challenge of 1990s, 27–28
Medical laboratory technician (MLT), 73–75
 certification of, 14
 in clinical laboratory
  roles/functions of, 35
  training of, 19
 education for, 35, 73–74
 requirements for, 74–75
Medical school, careers requiring, 110–111
Medical technologist examination, requirements for, 81
Medical technologist (MT), 76–81
 certification of, 80
 in clinical laboratory
  roles/functions of, 34
  training of, 19
 education for, 35, 76–79
  academic coursework in, 76
  programs for, 76–79
Medical technology (*see also* Clinical laboratory science (CLS))
 definition of, 2
Medicine, nuclear (*see also* Nuclear medicine)
Microbiology laboratory, 71–72
Military positions, employment opportunities in, 109–110
Morality, professional ethics and, 113–114

N

National Accrediting Agency for Clinical Laboratory Sciences (NAACLS), 55–56
National Certification Agency for Medical Laboratory Personnel (NCA), 53
National Registry in Clinical Chemistry (NRCC), 54–55
Nonmalfeasance, ethical behavior and, 114
Nontechnical health care-related professions, 105–106
Nonverbal communication, 127–128
Nuclear medicine, 83–86
 certification in, 85
 educational preparation for, 84–85
 job opportunities in, 84
 occupational description of, 84
 professional organizations for, 85–86
Nursing staff, public relations with, 125

O

Obligation, professional ethics and, 112
Organization(s)
   informal, of hospitals, 14–15
   professional, 57–62, 141–144 (*see also* Professional organizations)
Organizational communication systems, 130
Organization chart, hospital, 12–14
Ownership, hospital classification based on, 10

P

Pap smear, history of, 6
Partial thromboplastin time, activated (aPTT), 64–65
Pathologist in clinical laboratory, 17, 32–33
Patients
   public relations with, 125
   special, coping with, 128
   stay of, duration of, hospital classification based on, 10
Pediatric patients, communication with, 128
Personal characteristics, 22–23
Personnel
   future supply and demand for, 134–135
   licensure of, 45–48 (*see also* Licensure, personnel)
Phlebotomist in clinical laboratory, 19, 36–37
Physical requirements, 23
Physician, public relations with, 125
Prestige as challenge of 1990s, 28–29
Private laboratory, laboratory technical positions in, 103
Profession
   challenges to, of 1990s, 25–30
      CLS program closure as, 25–26
      declining student enrollment as, 25–26
      employee turnover as, 26–27
      image of profession as, 28–29
      increasing state and federal regulations as, 29–30
      in job satisfaction, 26–27
      manpower shortage as, 27–28
      new technology as, 30
      prestige as, 28–29
   image of, as challenge of 1990s, 28–29
   pros and cons of, 23–24
Professional ethics, 112–120
   AIDS and, 117–118
   Code of Ethics of American Society for Medical Technology and, 112–113
   confidentiality and, 116–117
   decision-making model in, 115–116
   morality and, 113–114
   obligation and, 112
   professionalism and, 112
   situations involving, 119–120
Professionalism, professional ethics and, 112
Professional organizations, 57–62, 141–144
   American Association for Clinical Chemistry (AACC) as, 59
   American Association of Bioanalysts as, 59
   American Association of Blood Banks (AABB) as, 59
   American Association of Immunologists (AAI) as, 60
   American Society for Medical Technology (ASMT) as, 57–58
   American Society for Microbiology (ASM) as, 61
   American Society of Allied Health Professions (ASAHP) as, 60
   American Society of Clinical Pathologists (ASCP) as, 58–59
   American Society of Hematology as, 60
   Clinical Laboratory Management Association (CLMA) as, 61
   College of American Pathologists (CAP) as, 62
   International Association of Medical Laboratory Technologists (IAMLT) as, 62
Professional service departments in hospitals, 15
Program closures as challenge of 1990s, 25–26
Progress in health care, 131
Prothrombin time (PT), 64–65
Public access, hospital classification based on, 10
Public health laboratory, laboratory technical positions in, 104
Public relations, 124–125

Q

Quality assurance (QA) programs, employment opportunities as, 105
Quality control, introduction of, 7

R

Radioimmunoassay, development of, 7
Registration, 44–45
Regulation, state and federal, increasing, as challenge of 1990s, 29–30
Research, employment opportunities in, 107

S

Safety directors, employment opportunities as, 106
Salaries, 37–39
Secretaries in clinical laboratory, 19

Section Supervisors in clinical laboratory, 17, 19
Senior Technologist in clinical laboratory, 19
Service, type of, hospital classification based on, 10–11
Situations, ethical, 119–120
Specialist in blood bank technology in clinical laboratory, 37
Staff, laboratory, 32–37
   clinical laboratory scientist or medical technologist on, 34–35
   clinical laboratory technician or medical laboratory technician on, 35
   clinical sciences specialists on, 33–34
   pathologist on, 32–33
   phlebotomist on, 36–37
   specialist in blood bank technology on, 37
State regulations, increasing, as challenge of 1990s, 29–30
Student enrollment, declining, as challenge of 1990s, 25–26
Supervision, employment opportunities in, 108

T

Teaching, hospital classification based on, 10
Technology(ies), new
   as challenge of 1990s, 30
   impact of, 133–134
Telephone manners, 129
Transplantation laboratory, laboratory technical positions in, 104

U

Uncooperative patients, communication with, 129
Urinalysis laboratory, 66
U.S. health care system, 9
Utilization review coordinators, employment opportunities as, 105

V

Vertical integration, hospital classification based on, 11
Veterinary medicine, laboratory technical positions in, 105
Veterinary school, careers requiring, 111

W

Waste control managers, employment opportunities as, 106